# SCIENCE
## ON THE WEB

**Springer**
*New York*
*Berlin*
*Heidelberg*
*Barcelona*
*Budapest*
*Hong Kong*
*London*
*Milan*
*Paris*
*Santa Clara*
*Singapore*
*Tokyo*

EDWARD J. RENEHAN, JR.

# SCIENCE
## ON THE WEB

*A Connoisseur's Guide to Over 500
of the Best, Most Useful, and
Most Fun Science Websites*

Springer

Edward J. Renehan, Jr.
58 Virginia Avenue
North Kingstown, RI 02852-6032 USA
EJR@ibm.net

Netscape is a registered trademark of Netscape Communiations Corporation.
Windows is a registered trademark of Microsoft Corporation.
Windows 95 is a registered trademark of Microsoft Corporation.
Windows NT is a registered trademark of Microsoft Corporation.
MS-DOS is a registered trademark of Microsoft Corporation.
Macintosh is a registered trademark of Apple Computer, Inc.
Java is a registered trademark of Sun Microsystems.

Cover illustration by Maura Condrick.

Library of Congress Cataloging-in-Publication Data
Renehan, Edward J.
     Science on the Web : a connoisseur's guide to over 500 of the
best, most useful, and most fun science Web sites / Edward J. Renehan.
       p.  cm.
     Includes bibliographical references and index.
     ISBN 0-387-94795-7 (softcover : alk. paper)
     1. Science–Computer network resources–Directories. 2. World
Wide Web (Information retrieval system)–Directories I. Title.
Q179.97.R46 1996
025.06'5–dc20                                  96-18688

Production managed by Natalie Johnson and Robert Wexler; manufacturing supervised by Joe
   Quatela.
Typeset in TEX from the author's Microsoft Word files.
Printed and bound by R.R. Donnelley & Sons, Harrisonburg, VA.
Printed in the United States of America.

9 8 7 6 5 4 3 2 1

ISBN 0-387-94795-7 Springer-Verlag New York Berlin Heidelberg     SPIN 10539297

**To John Perry Barlow,**

freedom fighter on the electronic frontier,

http://www.eff.org/~barlow

# CONTENTS

# INTRODUCTION

*Knowledge is of two kinds. We know a subject ourselves, or we know where we can find information upon it.*

**—Dr. Samuel Johnson**

Here is a quiz for you. What is more famous than Michael Jackson and more misunderstood than the economics of the federal budget deficit? What is the thing that, if it had been a person rather than an abstract combination of myriad circuits and wires, would have surely replaced Newt Gingrich as *Time*'s man of the year based on its profound informational (and therefore political, economic, and socioeconomic) impact in 1995? Why, the World Wide Web, of course—that dashing, precocious cyberchild of old man Internet. You remember the Internet, don't you?

Most scientists have been on familiar terms with the Internet for quite some time. The fabled network of networks was, after all, first created with scientists in mind. Up until a few years ago, the Internet was used almost exclusively by scientists, academics, and students as a vehicle for sharing information and research. In 1985 the Internet boasted only 1,961 host computers and numbered its users in the tens of thousands. But as many scientists know only too well, the once-pristine electronic frontier of the Internet has been overrun by new settlers in recent years. In fact, the Internet has doubled in size every ten months for the past six years. Today the number of users of the Internet increases at a rate of two million new logins each month—the equivalent of four new users every minute.

Along with vast numbers of "lay" users of the Internet have come vast numbers of purveyors of all sorts of information: the new gold of the electronic frontier toward which so many settlers rush. Scientific information certainly continues to have its place in the newly settled territories of the Internet. And scientific resources continue to grow in number. However, scientific information on the Net no longer constitutes the majority share. Scientific information plays second fiddle to financial information, religious information, erotic information, political information, literary information, and so on.

On the "plus" side, the information resources of the Internet seem to increase at almost as fast a rate as do the number of Internet users. The steadily growing profusion of information options—including information options related to the sciences—is wonderful. It is also utterly confusing.

# WHAT IN THE WORLD IS THE WORLD WIDE WEB?

The latest growth area for information resources on the Internet is the aforementioned World Wide Web. The "Web"—as it is familiarly called—provides access to virtually all those computers (servers) on the Internet that offer hypermedia-based information and documentation.

Hypermedia is a technology that presents and relates information by using nonlinear, nonsequential links rather than linear sequences. (Less formally put, hypermedia and hypertext enable users to navigate both the World Wide Web and the documents on it with point-and-click ease. To navigate, one "clicks" on words, phrases, and icons in a document, which provide links that enable you to jump at will to a new location in the document, or even to a new document altogether.) In short, the Web is a uniquely intuitive and information-rich environment.

Additionally, the Web is hospitable to graphic images, photographs, audio, and even full-motion video. Thus the Web has a multimedia capability that is of great value to scientists. Astronomers can view full-color space images on-line. Oceanographers can access real-time "remote sensor" data from key oceanographic sites around the world 24 hours a day. Students of chaos theory on the East Coast can connect and watch fractal trees generate on a minicomputer in Los Angeles. And paleontologists can get audio and image clips of Stephen Jay Gould giving a series of lectures at Harvard.

An additional perk of Web technology is that the Web provides easy tools for inexpensive on-line publication. Combining global connectivity and individual empowerment, the Web enables anyone who has a computer and the proper Internet connection to become a multimedia publisher. With the right tools (most of them available as free downloads from sites highlighted in this book) and a little effort, you can easily translate scientific papers into electronic Web documents (also known as "pages" or "sites") that the entire world can access. The same goes for reports, calls for papers, conference proceedings, announcements, course catalogs, etc. For more on this see the section of this book entitled "A Few Web Fundamentals/General Web Resources."

## WHERE DID THE WEB COME FROM?

Appropriately enough, the idea for the Web came from scientists—just as had the original idea for the Internet.

In 1989 Tim Berners-Lee, a physicist at the European Particle Physics Laboratory (CERN), proposed the concept of the Web as a system for transferring ideas and research among scientists in the high-energy-physics community.

Berners-Lee's original proposal defined a very simple implementation that used hypertext but did not include multimedia capabilities.

Something very much like this was introduced on Steve Jobs's NeXT computer system in 1990. The NeXT implementation allowed users to create, edit, view, and transmit hypertext documents over the Internet. The system was demonstrated for CERN committees and attendees at the Hypertext '91 conference.

In 1992 CERN began publicizing the World Wide Web (WWW) project and encouraging the development of Web servers at laboratories and academic institutions around the world. (In June 1993 there were only 100 WWW servers. Today, as I write, there are more than 15,000 with hundreds more coming on-line every day.) At the same time, CERN promoted the development of WWW clients (browsers) for a range of computer systems including X Windows (UNIX), the Apple Macintosh, and PC/Windows. (One of the most popular and useful of these has been MOSAIC, developed at the National Center for Supercomputing Applications and available free as a download from the NCSA Web site delineated elsewhere in this book.)

# THE RAISON D'ETRE FOR THIS BOOK

Today there are literally millions of documents on the Web. Every subject known to humankind can be found here. But finding *what you want* amid this mountain of data can be time-consuming.

Even if you use one of the popular search-engines on the Web to isolate all Web sites containing information on a given topic such as comets (or thermodynamics, or lipids, or polychlorinated byphenyls, etc.), you will still have to spend a fair amount of time browsing through the many documents called up by your search in order to see which ones provide the richest information base.

One comet page, for example, may contain nothing more than a collection of 200-year-old observations of Comet Halley with no accompanying links, while another page will provide a cornucopia of information and resources on all aspects of the study of comets from ancient times right up to the present day, with a long list of related links, including an appropriate description of (and a link to) the limited Comet Halley page for those who want it.

Which of these two documents is more useful? Which would you prefer to spend time with? The latter document is the one you will find discussed in this book.

In writing this book I have endeavored to provide a guide to the most useful and informationally rich resources for scientists on the Web. I have scoured the various Web information options in a range of scientific disciplines and cut out the shallow and trivial in favor of the deep and meaningful. Thus, what you have here is a directory to the most ambitious science pages on the Web, not only rich in links that leverage to the utmost the possibilities of hypertext but also rich in layers of vital, current data as represented in text, graphics, and audio.

# HOW TO USE THIS BOOK

The first section, "A Few Web Fundamentals/General Web Resources," gives you complete information on various options for connecting to the World Wide Web, including slip and PPP connections for private-home or small-office PCs and Macs as well as institutional connections via laboratories and universities. It also details the various hardware options available and includes a discussion of modems and baud rates—a vital aspect of Web usage since the graphics-intensive nature of the Web requires a fast modem connection. The chapter also provides details on the various software options for Web access from various platforms/modems, including not only browser software but also dialers and related utilities. Additionally, you will learn how to use the three primary search engines of the Web (WebCrawler, Lycos, and Yahoo) to find particular discipline resources on-line. And you will be introduced to the straightforward elements of creating your own Web documents.

The second and largest section provides listings and descriptions, by category, of what I consider to be the 500 most essential Web resources for scientists. The categories are alphabetized, from Artificial Intelligence to Zoology, as are the listings within each category. These listings include lab-specific Web pages as well as Web pages related to major academic departments, scientific journals and magazines, and scientific associations. Titles of documents are followed by the address location for the document (commonly known as the "URL" or the "http address"), and then by a description of the contents of the document. The structure of the book is designed to make it as easily navigable as the useful Web destinations that it endeavors to catalogue.

I have included, for each category discipline, at least one site that I call a "metasite." A metasite is a Web document that embraces within it virtually every other significant URL address within the discipline. The metasite for each category is denoted in the book by the logo in the margin at the beginning of this paragraph. Thus, although I have been highly selective in defining sites for inclusion here, I have also made a point of directing readers to intersections on the Web from which they will best be able to launch their own explorations and perhaps find emeralds that I did not.

Another logo that you should watch for when going through the book is the one in the margin by this line. This denotes a site that provides some signficant software or other valuable commodity for free. The commodity might be downloadable expert system shell software, downloadable images, a sample copy of a print journal, a free subscription to an electronic journal or magazine (a zine), or a downloadable HyperText Markup Language (HTML) editor. Make no mistake: access to the information on all the Web sites elucidated in this book is free. But this logo is reserved for sites that invite you take images, subscriptions, or software away with you.

Well, that about wraps it up. I suppose it is time to get on with the show. All I have left to say is that I certainly hope *Science on the Web* proves to be a useful research and study tool for you and your colleagues.

**Edward J. Renehan, Jr.**
Wickford, North Kingstown, Rhode Island, USA
EJRen@ibm.net
http://home.aol.com/EJRen

# A FEW WEB FUNDAMENTALS/ GENERAL WEB RESOURCES

This chapter briefly addresses two key questions related to accessing the World Wide Web: (1) How can I connect to the Web? (2) What tools (browsers, HTML-editors, search engines, etc.) are there available on-line to help me use the Web?

## HOW CAN I CONNECT TO THE WEB?

Time was when only a privileged few could get on the Internet superhighway and drive. That's not so anymore. These days, there are more and more "on-ramps" for the highway, and the tolls on the road are decreasing every day. Today you can speed around the Net and the Web for little more than the price of a subscription to the fruit-of-the-month club. Your connection options include permanent direct connections, dial-up connections to local hosts, and connection to the Web via a commercial on-line service.

## PERMANENT DIRECT CONNECTIONS: WEB NIRVANA

Those of us affiliated with universities, research labs, and corporations often have access to permanent direct connections. These are lines that connect directly to a TCP/IP (Transmission Control Protocol/Internet Protocol) network, which is in turn connected to the Internet. If you are at a university or large company, there is probably some kind of network connection hooked to your PC, Mac, or UNIX workstation, and you run TCP/IP support software by which you can move onto the Web and drive around the world. Bully for you. You have unlimited access and someone else is footing the bill.

Permanent direct connections are the very best way to travel the Web, as they allow fast data throughput capable of dealing swiftly with memory-fat Web graphics. You are doubly lucky because permanent direct connections are very expensive. They require dedicated high-speed telephone lines. The faster the line, the more expensive it is. And high-speed access is vital for many applications related to scientific research and communications (video conferencing), which require extremely fast transmissions of large amounts of data. Thus, not only are you driving for free, courtesy of your university or corporation, but you are driving a Mercedes.

Dedicated, full-time, high-speed connections such as these are absolutely fabulous. They not only enable people in an organization to access the Web, but also enable organizations to make their computers WWW servers. However, the annual fees for such speed and flexibility can easily exceed $10,000.

## LOCAL HOST DIAL-UP CONNECTIONS: THE POOR MAN'S WEB NIRVANA

Luckily, most people who want to connect with the Web simply don't need dedicated access. Home-based users, or small-office–based users, can easily use a modem (minimum 14.4 BPS [bits-per-second] recommended) to link his or her PC or Mac to a service provider's computer. These service providers are usually called *local hosts*. A local host computer runs with applications software that uses the TCP/IP protocols to communicate with other Internet host computers directly. And the type of access you are granted to the Web, via your local host, is called *dial-up access*.

To communicate with the Web via your local host, you must use software that enables your computer to use the TCP/IP language to communicate over local telephone lines. Here you have two choices. The first is SLIP (Serial Line Internet Protocol) and the second and newer option is PPP (Point-to-Point Protocol). These low-cost alternatives provide full peer access to the Internet. The difference between the two is fundamental. SLIP does not provide error correction or data compression, but it still works well for home and small-business applications. PPP was specifically developed to rectify SLIP's error-correcting weakness. PPP checks incoming data and asks the sending computer to retransmit when it detects an error in an IP packet. Thus, of the two protocols, I recommend PPP. It'll save you time.

A few popular providers of local host dial-up connection are MCI (800-955-6505), the Pipeline (212-267-3636), and NETCOM (800-501-8649). I use IBM Net (800-888-4103). All of these outfits have proprietary SLIP or PPP software packages, complete with browsers, that they'll be happy to send to you and with which you can connect to their services.

## CONNECTION TO THE WEB VIA COMMERCIAL ON-LINE SERVICES: THE VERY POOR MAN'S WEB NIRVANA

The major commercial on-line services now offer Web access. The two leaders in the field here are America Online (800-827-6364) and CompuServe (800-554-4067 or 800-848-8199). Each service provides you with its own Web browser software and offers both 14.4 BPS and 28.8 BPS access. Commercial on-line services provide the least expensive Web access going. But there are trade-offs. Load times can often be very slow, depending on the volume of people on the network at any given time. And the proprietary browsers used by the services are not great: images can look pretty bad. However, depending on how much time you have and how important graphics are to you, these on-line services can provide economical and easy access to the Web.

# WHAT TOOLS ARE AVAILABLE TO HELP ME USE THE WEB?

## BROWSER TOOLS

As mentioned earlier, on-line services such as America Online and CompuServe provide their own browser software, as do many dial-up local hosts. While these proprietary browsers are okay for starters, you will probably want to graduate to a first-rank, stand-alone commercial browser in short order. And the browser I have in mind is Netscape, the best commercial implementation of Mosaic to be had. Netscape loads files faster than any other browser, and delivers a level of graphical clarity second to none. It also supports in-line JPEG (Joint Photographic Experts Group) pictures and—with the appropriate extensions—even Java-style 3D. As a further convenience, Netscape lets you participate in newsgroups and send e-mail within the Netscape browser window, and launch software applications while on the Web.

To download Netscape as well as other browsers, including that old standby NCSA Mosaic, you can simply go to one site on the Web and find absolutely everything you need, including versions of Netscape for PCs, Macintosh computers, and even UNIX work-stations. The site is entitled *Internet Tools, Browsers and Viewers* and the address is http://metro.turnpike.net/Rene/tools.htm. Once you download Netscape and install it, you can go to another handy Web site to enjoy a great hypertext tutorial on how to use Netscape to its fullest. This site is entitled *Netscape Tutorial* and the address is http://w3.ag.uiuc.edu/AIM/Discovery/Net/www/netscape/index.html.

# GENERAL INTERNET/WEB TUTORIALS ON-LINE

There are a number of Web pages that provide excellent hypertext tutorial instruction on the ins and outs of the Internet and the Web. Here is a quick listing of a few of the best of them.

Easy Internet

**http://www.futurenet.co.uk/netmag/Issue1/Easy/index.html**

The Electronic Frontier Foundation's Extended Guide to the Internet

**http://www.eff.org/papers.bdgtti/eegtti.html**

The Global Village Internet Tour

**http://www.globalcenter.net/gcweb/tour.html**

The Glossary of Internet Terms

**http://www.matisse.net/files/glossary.html**

How to Publish on the Web

**http://www.thegiim.org/**

Imajika's Guide for New Users

**http://www.sjr.com/sjr/www/bs/**

Internet E-mail Syntax Guide
**http://starbase.nse.com/~shoppe/mailgd4.htm**

Internet Exploration Using Mosaic
**http://www.math.udel.edu/MathResources.html**

Internet Foreplay
**http://www.easynet.co.uk/pages/forepl/forepl.html**

MecklerMedia Web Guide
**http://www.mecklerweb.com/webguide/entry.htm**

Winsock: A Beginner's Guide
**http://sage.cc.purdue.edu/~xniu/winsock.html**

# HTML EDITORS FOR CREATING YOUR OWN WEB DOCUMENTS AND PAGES

Many readers of this book will want not only to read what others have published on the Web, but also to do some Web publishing themselves.

The tool that you use to create hypertext documents for the World Wide Web is called HyperText Markup Language (HTML). If you want to create your own homepage, or render a document in a form readable on the Web, you need to get a good HTML editor and learn how to use it. This involves assigning document tags and working with basic text structures. You may also want to learn how to incorporate images into your HTML documents.

You'll be glad to hear that HTML is not all that hard to master. One of the best introductions to HTML that I've found is freely available on the Web itself and is entitled *Introduction to HTML* (http://www.cwru.edu/help/introHTML/toc.html).

Before you can learn to use an HTML editor, however, you have to have one available on your platform. Many excellent HTML editors are available on the Web for Macintosh, Windows, and UNIX machines. Here are some addresses where you'll find them available for download:

## PC/WINDOWS HTML TOOLS

Ant HTML and Ant PLUS—Works with Word for Windows, NT, and Windows 95 (and with International versions of Word 6.0 and above), and supports any and all HTML tags.

**http://mcia.com/ant**

Gomer HTML Editor for Windows

**http://clever.net/gomer**

Hot Dog Web Editor—A fast, flexible, and friendly HTML editor for Windows.

**http://www.sausage.com/**

HTML Easy! Pro—For Microsoft Windows, Windows 95, and NT, this editor supports colors and texture, HTML 3.0, and Netscape extensions.

**http://www.seed.net.ts/˜milkylin**

HTML Writer—A Microsoft Windows applications designed to simplify the creation or editing of HTML documents.

**http://lal.cs.byu/people/nosack**

HTMLed—An HTML editor for Microsoft Windows.

**ftp://tenb.mta.ca/pub/HTMLed/**

Internet List Keeper—A very simple, template-based HTML tool for creating Web pages.

**http://www.drweb.com/**

Kenn Nesbitt's WebEdit—This Windows-based HTML editor includes support for every feature of HTML 3.0, plus Netscape extensions.

**http://www.nesbitt.com**

TILE—This great Web authoring program for Lotus Notes works best for databases and collaborative authoring.

**http://tile.net/info/viewlist.html**

Web Media Publisher—A full-featured 32-bit HTML editor with complete Java and Shockwave support.

**http://www.wbmedia.com/software.html**

Web Publisher—This automated production tool for the Web and HTML allows you to seamlessly convert and enhance documents originated with standard word processing software.

**http://www.skisoft.com/skisoft/**

Webber—This editor features SGML (Standard Generalized Markup Language) compliant validation and supports long file names as well as all tags for HTML versions 2.0 and 3.0, Netscape, and IExplorer.

**http://www.csdcorp.com/webber.htm**

WebElite—This is a quick, easy-to-use editor for Windows 3.1 and 95 that combines powerful features with low system requirements.

**http://www.safety.net/webelite**

WordPerfect Internet Publisher—This free add-on to WordPerfect 6.0 for Windows allows you to create HTML documents readable by Netscape and other browsers.

**http://wp.novell.com/elecpub/intpub.htm**

## MACINTOSH HTML TOOLS

Alpha Text HTML Editor for Macintosh

**http://www.cs.umd.edu/~keleher/alpha.html**

Ant HTML and Ant PLUS—Works with Word for Macintosh and supports any and all HTML tags.

**http://mcia.com/ant**

Arachnid—This software offers WYSIWYG design capabilities, can build a page from an existing HTML file, and can import fully formatted TRF text files to allow the use of text processors.

**http://sec-look.uiowa.edu**

BBEdit HTML Extensions

**http://www.uji.es/bbedit-html-extensions.html**

BBEDIT HTML Tools

**http://www.york.sc.uk/~ld11/BBEditTools.html**

GT HTML.DOT—Here are great Macintosh macros that provide a pseudo-WYSIWYG authoring environment for HTML.

**http://www.gatech.edu/word_html/**

HTML Grinder—This powerful utility for Macintosh-based Webmasters uses plug-in tools that perform repetitive chores on all of your HTML files at once.

**http://dragon.acadiau.ca/~giles/HTML_Editor/**
**Documentation.html**

HTML.edit—Another Mac-based editor for HTML.

**http://ogopogo.nhc.edu/tools/HTMLedit/HTMLedit.html**

Web Weaver—A Macintosh text-editor for creating and editing HTML documents quickly and easily.

**http://www.northnet.org/best/**

WebDoor—A Mac-based automated Web publishing system requiring no knowledge of HTML.

**http://www.opendoor.com/webdoor/**

## UNIX HTML TOOLS

ASHE—This stands for A Simple HTML Editor.

**ftp://ftp.cs.rpi,edu/pub/puninj/ASHE/README.html**

asWedit—A context-sensitive HTML 3.0 and HTML 2.0 editor with a built-in parser and a comprehensive "Help" system.

**http://sunsite.doc.ic.ac.uk/packages/www/asWedit/**

BullDozer WYSIWYG HTML Editing

**http://cscsun1.larc.nasa.gov/~rboykin/Dozen/**

City University HTML Editor—From London comes an HTML editor for X Windows using the Andrew Toolkit.

**http://web.cs.city.ac.uk/homes/njw/htmltext/htmltext.html**

Cyberleaf—Software for converting Microsoft Word, WordPerfect, Framemaker and Interleaf documents into HTML.

**http://www.ileaf.com/ip.html**

HTMLJive—Yes! An HTML editor written in Java-script! Create your own Web page incorporating manipulative 3D graphics that leverage VRML and Hot Java!

**http://www.cris.com/~raydaly/htmljive.html**

tkHTML: tcl/tk HTML Editor—This is a simple HTML editor based on the tcl script language and the tk toolkit.

**http://www.ssc.com/~roland/tkHTML/tkHTML.html**

## SEARCH ENGINES ON THE WEB

This book should be your first stop when searching for scientific resources on the Web. Even if you do not find a specific site listed within a discipline that sounds like it will meet your information needs, you will most assuredly find something at the metasites included here.

There is at least one metasite for each scientific discipline. These sites are hypertext documents that exist primarily to be intersections for as many links as possible in any one given scientific category. To help you find the metasites in this book, I have flagged them with their own logo, as follows:

Should the resources of a metasite not meet your needs, then there are a number of Web-based search engines that you should investigate. They are all available 24 hours a day on the Web for the mere price of oxygen. In other words, they are free. And they all operate very simply. Just enter some keywords for a search, survey the results, and click on those hyperlinks you'd like to follow.

The best search engines are as follows:

ALIWEB—Europe

**http://web.nexor.co.uk/public/aliweb/aliweb.html**

USA

**http://www.cs.indiana.edu/aliweb/search**

CUI Search Catalog

**http://cuiwww.unige.ch/w3catalog**

EINet Galaxy

**http://galaxy.einet.net/www/www.html**

Global Network Academy Meta-Library

**http://uu.gna.mit.edu:8001/uu-gna/meta-library/index.html**

Lycos

**http://lycos.cs.cmu.edu/**

Nomad Gateway

**http://www.rns.com/cri-bin/nomad**

WebCrawler Searcher

**http://www.webcrawler.com**

WWWW—The World Wide Web Worm

**http://www.cs.colorado.edu/home/mcbrian/WWWW.html**

Yahoo

**http://www.yahoo.com**

These are all quite good options. My two favorites are WebCrawler and Yahoo.

WebCrawler was developed by Brian Pinkerton at the University of Washington. Given a set of parameters, WebCrawler creates indexes of documents it locates on the Web and then lets you search the indexes. WebCrawler presents results in a prioritized order. The first item listed is the one with the highest number of "hits" within the text of the page for the keywords you have specified. Links noted at the bottom of a set of WebCrawler results are thus less likely to be what you are looking for than items at the top.

Yahoo is an extensive database of Web sites. There are more than 30,000 entries at the moment with 100 to 200 new links added each day. Yahoo's various pages are organized in subject categories, and you have the option of searching within a given category or to search the entire database. Once again, as with Webcrawler, a prioritized list of hyperlinks appears as the result of your search. To follow the links, just click.

Because of subtle differences in the manner in which they search the Web, WebCrawler and Yahoo can, at times, produce remarkably dissimilar results for the same set of parameters. A site that one misses, however, the other is sure to catch. That's why I recommend using both WebCrawler and Yahoo in tandem to ferret out the Web information you need.

# ARTIFICIAL INTELLIGENCE

The American Association for Artficial Intelligence

Artificial Intelligence Meta-Site

Association for Computing Machinery (ACM) Special
      Interest Group for Artificial Intelligence (SIGART)

Canon Natural Language Processing Group

Carnegie-Mellon University (CMU) Artificial Intelligence
      Repository

University of Chicago Artificial Intelligence Lab

The Computation and Language E-Print Archive

European Association for Logic, Language and Information

Free Compilers & Interpreters (including LISP)

Freeware Expert System Shells

The Journal of Experimental and Theoretical Artificial
      Intelligence (JETAI)

Harvard Robotics Lab

Iowa State University Artificial Intelligence Research Group

The Journal of Artificial Intelligence Research (JAIR)

Knowledge Representation Meta-Site

KQML: Knowledge Query and Manipulation Language

University of London: Queen Mary & Westfield College
      Distributed Artificial Intelligence Research Unit

The Maine Cooperative Distributed Problem Solving
      Research Group

MIT Artificial Intelligence Laboratory

The University of Massachusetts Distributed Artificial
      Intelligence Laboratory

The University of Massachusetts Laboratory for Perceptual
      Robotics

University of Michigan Distributed Intelligent Agents Group

Neurality: The Neural Networks Meta-Site

# AMERICAN ASSOCIATION FOR ARTIFICIAL INTELLIGENCE

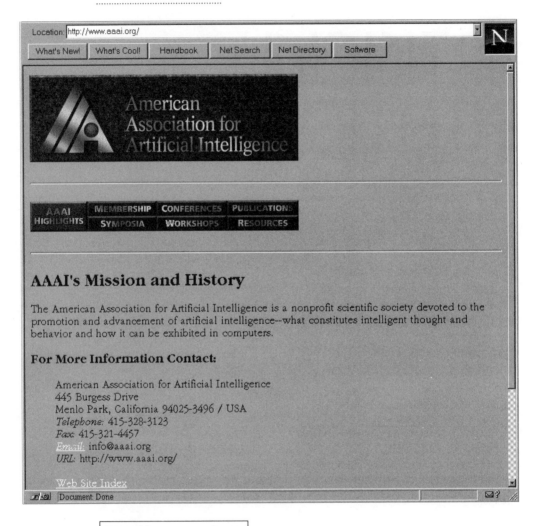

**http://www.aaai.org**

The nonprofit American Association for Artificial Intelligence (AAAI) is a scientific society devoted to the promotion and advancement of the science of artificial intelligence.

Launched in March 1995, the AAAI's web site now contains nearly a thousand files with more being added every week. Come to the AAAI web site for:

☐ membership information and applications, AAAI's Executive Council, current AAAI officers and staff members, and a complete list of AAAI fellows together with e-mail and snail-mail contact info;

☐ details on the full range of AAAI programs and offerings for members and nonmembers alike, including conferences, symposia and workshops that include the National Conference on Artificial Intelligence, the Innovative Applications of Artificial Intelligence Conferences, and the Knowledge Discovery and Data Mining Conference;

☐ information on publications ranging from *AI Magazine* and the *AI Directory*, technical reports, proceedings, and complete tables of contents (TOCs) for AAAI books distributed internationally by the MIT Press (with full text for some titles coming in the near future).

An especially useful feature of this site is the inclusion of author kits and guidelines (including templates and macros) for submitting papers to the various AAAI conferences and symposia. Also be sure to check out the AAAI's AI Educational Repository, a central registry of and distribution point for vital resources related to AI and education on the Web.

# ARTIFICIAL INTELLIGENCE META-SITE

**http://ai.iit.nrc.ca/ai_point.html**

Maintained by Canada's Institute for Information Technology, this Web page provides links to literally hundreds of Web resources related to artificial intelligence. Here you have pointers to AI bibliographies, publishers and journals, calls for papers for most upcoming AI conferences, job postings for AI workers, links to corporate home pages for AI-related companies, and much more. This meta-site should be in the browser hotlist for all Web surfers with an interest in AI.

# ASSOCIATION FOR COMPUTING MACHINERY (ACM) SPECIAL INTEREST GROUP FOR ARTIFICIAL INTELLIGENCE (SIGART)

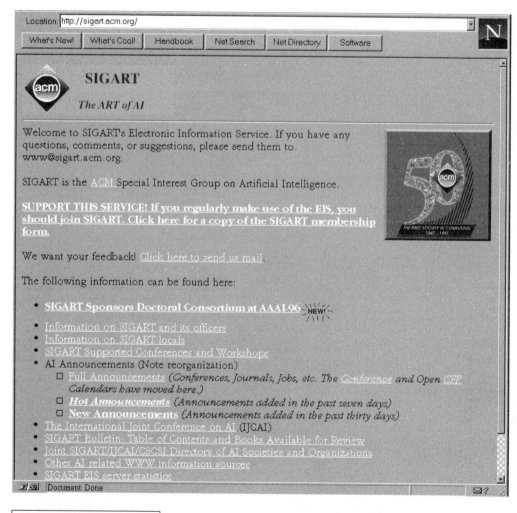

**http://sigart.acm.org**

Come to this Web document for complete information on the special programs of the ACM geared to meet the information needs of artificial intelligence professionals. Point and click to easily access details concerning SIGART, its officers (including e-mail addresses), local chapters, conferences, and workshops. Regularly updated announcement boards include listings of AI-related job openings. And additional pages of the SIGART document provide information on the SIGART Bulletin (TOCs and author guidelines), The International Joint Conference on AI, and how to join SIGART. There is also a handy on-line directory of AI societies and organizations, as well as selected links to AI-related Web information sources.

# CANON NATURAL LANGUAGE PROCESSING GROUP

http://www.cre.canon.co.uk/nlp.html

Canon/UK's Natural Language Processing Group was founded in 1990 and currently has a staff of seven. The group is engaged in long-term research on natural language software in a number of human languages for a variety of applications. Come to this Web site for details on:

❏ the EMMA project, which, like its predecessor NLI, involves getting machines to have robust understanding of ordinary human language (i.e., instant translation of speech and text). EMMA exploits contextual knowledge to get the best results for generic spoken dialogue systems using hypothetical reasoning, incremental fast-and-loose parsing and generation, and speech-understanding based on articulatory features;

❏ the ADROIT project, which involves the translation of Canon documentation (user manuals, etc.) into all the languages of all the European countries in which they are needed (typically about ten languages);

❏ job openings at Canon Research Engineering.

# CARNEGIE-MELLON UNIVERSITY (CMU) ARTIFICIAL INTELLIGENCE REPOSITORY

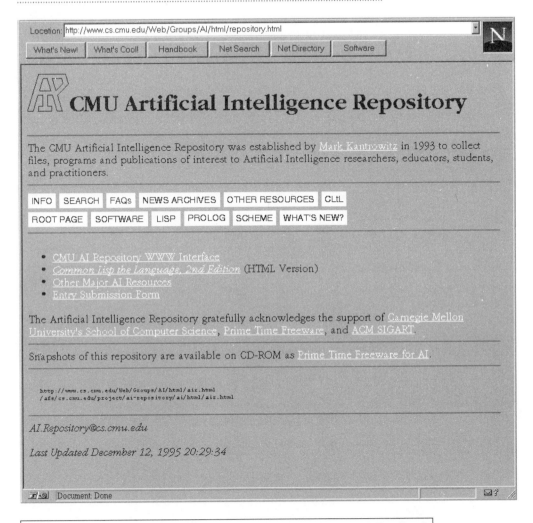

Location: http://www.cs.cmu.edu/Web/Groups/AI/html/repository.html

What's New! | What's Cool! | Handbook | Net Search | Net Directory | Software

## CMU Artificial Intelligence Repository

The CMU Artificial Intelligence Repository was established by Mark Kantrowitz in 1993 to collect files, programs and publications of interest to Artificial Intelligence researchers, educators, students, and practitioners.

INFO | SEARCH | FAQs | NEWS ARCHIVES | OTHER RESOURCES | CLtL
ROOT PAGE | SOFTWARE | LISP | PROLOG | SCHEME | WHAT'S NEW?

- CMU AI Repository WWW Interface
- Common Lisp the Language, 2nd Edition (HTML Version)
- Other Major AI Resources
- Entry Submission Form

The Artificial Intelligence Repository gratefully acknowledges the support of Carnegie Mellon University's School of Computer Science, Prime Time Freeware, and ACM SIGART.

Snapshots of this repository are available on CD-ROM as Prime Time Freeware for AI.

```
http://www.cs.cmu.edu/Web/Groups/AI/html/air.html
/afs/cs.cmu.edu/project/ai-repository/ai/html/air.html
```

AI.Repository@cs.cmu.edu

Last Updated December 12, 1995 20:29:34

Document: Done

## http://www.cs.cmu.edu/Web/Groups/AI/html/repository.html

The Artificial Intelligence Repository at Carnegie-Mellon was founded by Mark Kantrowitz in 1993 to provide an electronic clearing house for files, programs, and publications of interest to AI researchers, educators, students, and practitioners. Of special interest here is a complete, unabridged, HTML edition of Guy Steele's classic text *Common Lisp the Language*, second edition. Yes, we're talking about the same book you just paid $40 for at the university bookstore.

# UNIVERSITY OF CHICAGO ARTIFICIAL INTELLIGENCE LAB

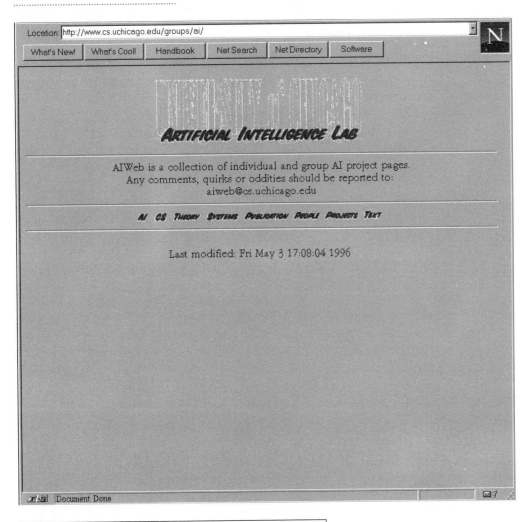

**http://www.cs.uchicago.edu/html/groups/ai/**

The faculty of the University of Chicago's AI Lab includes several of the great names in contemporary AI research, including Kristian Hammond, R. James Firby, Charles Martin, and Michael Swain.

The Lab's web site provides detailed descriptions of several of the Lab's ongoing research and development projects including:

❑ research on developing animate agents (i.e., creating intelligent, goal-directed behavior in software and hardware agents);

❑ CADI, an intelligent tutoring system that teaches medical students how to listen to hearts and diagnose various heart diseases based on what they hear;

❑ FindMe Systems, which are designed to help people navigate through vast spaces of information such as movie, automobile, and restaurant databases;

❑ Perseus, the architecture for the Lab robot's visual system, which recognizes when people are pointing and finds the object being pointed to;

❑ ROENTGEN, a case-based assistant for creating radiation therapy plans to treat cancer.

There is fun stuff here, too. The LAB CAM remote sensor lets you take a look around the University of Chicago's animate agent lab, see what Mike Swain is having for lunch, who is working and who isn't (get back to work, Michael), and so on. And then there is CyberCHEF, a case-based recipe retrieval system that is designed to respond to requests for recipes posted to the rec.food.recipes newsgroup.

# THE COMPUTATION AND LANGUAGE E-PRINT ARCHIVE

http://xxx.lanl.gov/cmp-lg/

The Computation and Language E-Print Archive is a fully automated archive and distribution server for papers on computational linguistics, natural-language processing, speech processing, and related fields.

Search by title, author, or both, and then retrieve abstracts or complete texts of papers from proceedings that include the 32nd annual meeting of the Association for Computational Linguistics (ACL '94), Proceedings of the ACL '94 Post-Meeting Workshops, the Third International Workshop on Tree Adjoining Grammars and Related Formalisms, the Fifteenth International Conference on Computational Linguistics (COLING '94), the 7th Conference of the European Chapter of the Association for Computational Linguistics (EACL '95) and the ACL SIGDAT Workshop, and the 33rd Annual Meeting of the Association for Computational Linguistics (ACL '95).

Useful links at this site connect you to the home pages for the Association for Computational Linguistics, the NLP/CL Universe (a comprehensive database of Web-accessible material related to computational linguistics and natural language processing), *Colibri* (an electronic newsletter on language, speech, logic, and information published at Utrecht University), and *Tal*, a Dutch students journal for Computational Linguistics that includes interviews with researchers (CL's first fanzine!)

# EUROPEAN ASSOCIATION FOR LOGIC, LANGUAGE, AND INFORMATION

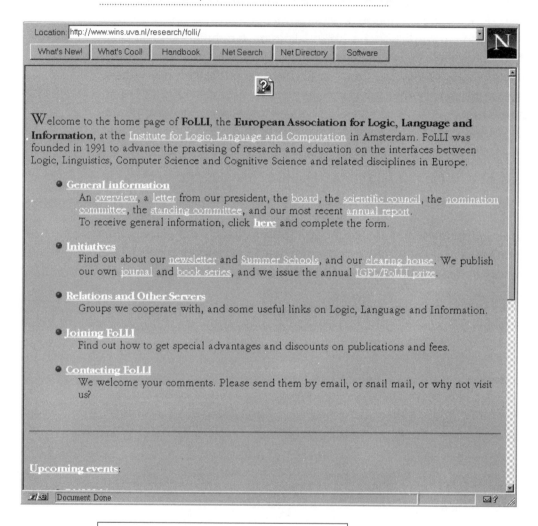

http://www.fwi.uva.nl/research/folli

Maintained by the splendid people at Amsterdam's Institute for Logic, Language, and Computation, this Web site provides details on the activities of the European Association of Logic, Language, and Information.

The organization was first conceived in 1990 by the prominent semanticist Martin Stokhof. His idea was that the exciting but still sparsely inhabited interface between logic, linguistics, and computer science needed some visible platform. The resulting umbrella organization managed to gather several enterprises under its aegis, including the Amsterdam Colloquia in Formal Semantics, the London-based Interest Group in Pure and Applied Logic (IGPL), and more conspicuously, the European Summer Schools in Logic Language and Information, which will enjoy their tenth season in 1997.

Perhaps mindful of Lenin's observation that a political party is a mob with a newspaper, the European Association for Logic, Language, and Information soon legitimized itself with an excellent journal, published by Kluwer. Gradually other activities were added, such as the noted TEMPUS program for academic exchange with Central and Eastern Europe. Complete information on all these activities can be found online.

# FREE COMPILERS AND INTERPRETERS (INCLUDING LISP)

http://cuiwww.unige.ch/cgi-bin/freecomp

Come to this Web site to find a searchable version of the wonderful Free Compilers List maintained by David Muir Sharnoff. The list catalogues free available software for language tools, which includes compilers, compiler generators, interpreters, and assemblers—things whose user interface is language. Here you will not only find numerous LISP compilers available for free downloading, but also natural language tools and a variety of robust Fortran and C++ implementations as well.

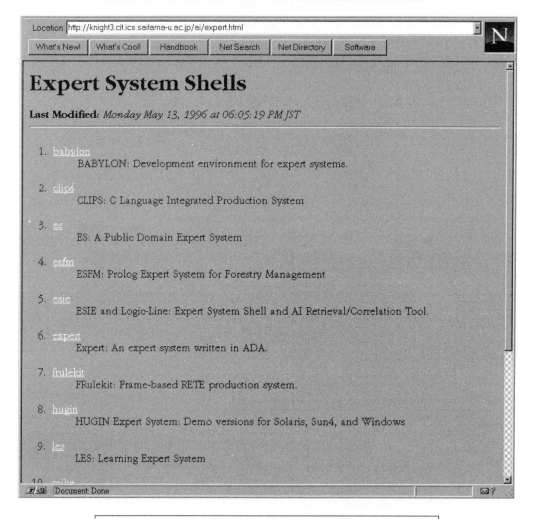

Location: http://knight3.cit.ics.saitama-u.ac.jp/ai/expert.html

| What's New! | What's Cool! | Handbook | Net Search | Net Directory | Software |

## Expert System Shells

**Last Modified:** *Monday May 13, 1996 at 06:05:19 PM JST*

1. babylon
   BABYLON: Development environment for expert systems.

2. clips
   CLIPS: C Language Integrated Production System

3. es
   ES: A Public Domain Expert System

4. esfm
   ESFM: Prolog Expert System for Forestry Management

5. esie
   ESIE and Logic-Line: Expert System Shell and AI Retrieval/Correlation Tool.

6. expert
   Expert: An expert system written in ADA.

7. frulekit
   FRulekit: Frame-based RETE production system.

8. hugin
   HUGIN Expert System: Demo versions for Solaris, Sun4, and Windows

9. les
   LES: Learning Expert System

10. mike

Document: Done

**http://knight3.cit.ics.saitama-u.ac.jp/ai/expert.html**

Here is the place to come for twelve great expert system shells, each of them available for you to download for free. The offerings include:

❏ BABYLON: Development environment for expert systems

❏ CLIPS: a C-language integrated production system

- ES: a public domain expert system

- ESFM: a Prolog expert system for forestry management

- ESIE and Logic-Line: an expert system shell and AI retrieval/correlation tool

- Expert: an expert system written in ADA

- FRulekit: a frame-based RETE production system

- HUGIN Expert System in demo versions for Solaris, Sun4, and Windows

- LES: Learning Expert System

- MIKE: a great portable expert system teaching system

- OPS5: a RETE-based expert system shell

- PROTEST: a Prolog expert system building tool.

# HARVARD ROBOTICS LAB

http://hrl.harvard.edu

The Harvard Robotics Laboratory was founded in 1993 by Roger Brockett (Wang Professor of Computer Science and Electrical Engineering). The lab's current projects include research in computational vision, neural networks, tactile sensing, motion control, and VLSI systems.

The Web page of the Harvard Robotics Lab provides the full text of a number of major papers authored or coauthored by Brockett, including papers concerning hybrid models for motion control systems, differential equations and matrix inequalities on isospectral families,

dynamical systems and their associated automata, pattern generation and the control of nonholonomic systems, and the dynamics
of kinetic chains. The site also provides details on:

❏ The vision research of Professors David Mumford and Alan Yuille,
who are working to construct more and more inclusive probability models for the set of all possible images, and to use these
together with Bayesian probability theory to find estimates of the
3D structure behind real-world 2D images;

❏ Professor Robert Howe's ongoing research concerning the biome-
chanics of the human hand and the related modeling of
mechanoreceptors, wherein he is working to elucidate impe-
dance selection in dexterous manipulation;

❏ And work by other researchers on such topics as deformable
templates for object recognition, genetic algorithms and the min-
imization of 2D energy functionals, neural network models of
visual perception, and the role of pule-like waveforms in the
design of dynamic systems that simulate finite state machines.

Additional reports available via the Harvard Robotics Lab Web site
include details on a project to determine the sensing and control
strategies that will permit a legged robot to carry a payload smoothly
over rough ground, the development of new tactile sensors and the
integration of sensed information with real-time control, and the development of new display devices that allow a human operator to
literally feel the objects that a remote robot is handling.

# IOWA STATE UNIVERSITY ARTIFICIAL INTELLIGENCE RESEARCH GROUP

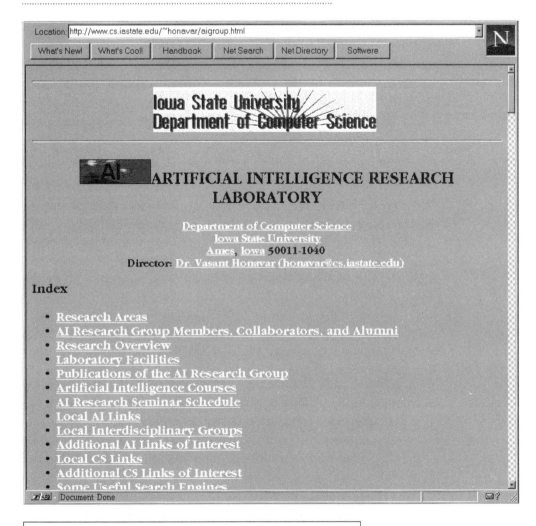

Location: http://www.cs.iastate.edu/~honavar/aigroup.html

| What's New! | What's Cool! | Handbook | Net Search | Net Directory | Software |

**Iowa State University Department of Computer Science**

**ARTIFICIAL INTELLIGENCE RESEARCH LABORATORY**

Department of Computer Science
Iowa State University
Ames, Iowa 50011-1040
Director: Dr. Vasant Honavar (honavar@cs.iastate.edu)

**Index**

- Research Areas
- AI Research Group Members, Collaborators, and Alumni
- Research Overview
- Laboratory Facilities
- Publications of the AI Research Group
- Artificial Intelligence Courses
- AI Research Seminar Schedule
- Local AI Links
- Local Interdisciplinary Groups
- Additional AI Links of Interest
- Local CS Links
- Additional CS Links of Interest
- Some Useful Search Engines

Document: Done

**http://www.cs.iastate.edu/~honavar/aigroup.html**

Directed by Dr. Vasant Honaver, the Artificial Intelligence Research Group at Iowa State is focused on the design and systematic exploration of a family of algorithms for incremental construction of neural networks for pattern classification and related applications. This work emphasizes the characterization and use of alternative representational and inductive biases (e.g., architectural constraints) for rapid construction of a near-minimal network for a given problem.

Related research by the group in this area seeks to automate the exploration of the design space of neural networks using genetic algorithms. Of particular interest here are efficient (artificial) genetic representations of neural architectures for perception and action in autonomous agents and robots. Other work on machine learning has focused on efficient algorithms for induction or regular grammars from examples and queries. Techniques for induction of tree grammars, attributed grammars, and logic programs are also under investigation.

The group's Web site includes electronic texts of numerous research reports and other publications generated by members on such topics as parallel and distributed architectures for knowledge representations and reasoning (focusing on fault-tolerant neural network architectures for data storage and retrieval, syntax analysis, and inference). Additional papers highlight how recent research on temporal reasoning has led to the development of a new framework—stochastic temporal constraint networks—for reasoning about events in time under stochastic uncertainty.

Additional topics include the application of AI to adaptive self-managing communication networks, multimedia data and knowledge base management, engineering design, robotics, adaptive heuristics for routing in large communication networks, and computational investigations of mechanisms underlying neural plasticity, learning, and memory.

# JOURNAL OF ARTIFICIAL INTELLIGENCE RESEARCH (JAIR)

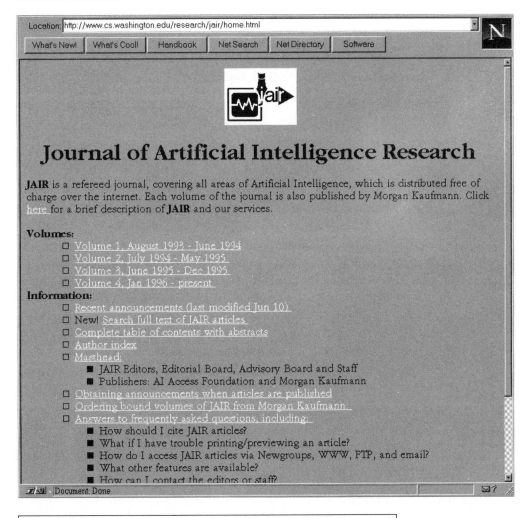

Location: http://www.cs.washington.edu/research/jair/home.html

| What's New! | What's Cool! | Handbook | Net Search | Net Directory | Software |

## Journal of Artificial Intelligence Research

**JAIR** is a refereed journal, covering all areas of Artificial Intelligence, which is distributed free of charge over the internet. Each volume of the journal is also published by Morgan Kaufmann. Click here for a brief description of **JAIR** and our services.

**Volumes:**
- Volume 1, August 1993 - June 1994
- Volume 2, July 1994 - May 1995
- Volume 3, June 1995 - Dec 1995
- Volume 4, Jan 1996 - present

**Information:**
- Recent announcements (last modified Jun 10)
- New! Search full text of JAIR articles
- Complete table of contents with abstracts
- Author index
- Masthead:
  - JAIR Editors, Editorial Board, Advisory Board and Staff
  - Publishers: AI Access Foundation and Morgan Kaufmann
- Obtaining announcements when articles are published
- Ordering bound volumes of JAIR from Morgan Kaufmann:
- Answers to frequently asked questions, including:
  - How should I cite JAIR articles?
  - What if I have trouble printing/previewing an article?
  - How do I access JAIR articles via Newgroups, WWW, FTP, and email?
  - What other features are available?
  - How can I contact the editors or staff?

Document: Done

## http://www.cs.washington.edu/research/jair/home.html

The *Journal of Artificial Intelligence Research* (JAIR) is a free electronic journal distributed over the Internet by the nonprofit AI Access Foundation. (Note: A print-edition of JAIR is published simultaneously by Morgan Kaufmann, and is not free.)

Use this Web site to enroll for your free subscription to the electronic edition of JAIR, to search the full texts of JAIR articles dating back to August 1993, to access issue TOCs with abstracts, and to learn how to submit papers to JAIR.

Among the dozens of useful articles you will find here are N. Nilsson on teleo-reactive programs for agent control, M.L. Ginsberg on dynamic backtracking, and M. Buchheit et al. on decidable reasoning in terminological knowledge representation systems.

# THE JOURNAL OF EXPERIMENTAL AND THEORETICAL ARTIFICIAL INTELLIGENCE (JETAI)

http://turing.paccs.binghamton.edu/jetai

The purpose of the *Journal of Experimental and Theoretical Intelligence* (JETAI) is to advance scientific research in AI by providing a public forum for the presentation, evaluation, and criticism of research results, the discussion of methodological issues, and the communication of positions, preliminary findings, and research directions. The scope of the journal encompasses all subfields of AI research including problem solving, perception, learning, knowledge representation and memory, and neural system modeling.

JETAI is edited by Eric Dietrich, professor of philosophy and cognitive science at SUNY Binghamton, and is published by Taylor & Francis.

Come to this Web site for author guidelines, and for information about JETAI's international editorial advisory board. Additionally, an easy point-and-click interface lets you search through the contents, abstracts, and editorials of the journal's past editions. You may even request, via e-mail, a free trial issue.

# KNOWLEDGE REPRESENTATION META-SITE

**http://mnemosyne.itc.it:1024/kr-links.html**

Here is one-stop shopping for all the most vital Web links related to knowledge representation. You have more than sixty options here, among them Web pages focusing on:

❏ Ontology—projects, people, conferences, and specific resources;

❏ Mechanized Deduction—a service for the mechanized reasoning community;

❏ Formal Methods—the home page of Formal Specification Languages;

❏ Compulog Net—ESPRIT's Network of Excellence on computational logic;

❏ The Association for Automated Reasoning;

❏ ECCC—The Electronic Colloquium on Computational Complexity;

❏ Z Notation—the formal specification notations Z, based on set theory and first-order predicate logic;

❏ The Logics Workbench—an interactive theorem prover created at the University of Berne, Switzerland;

❏ Numerous national and international conferences.

# KQML: KNOWLEDGE QUERY AND MANIPULATION LANGUAGE

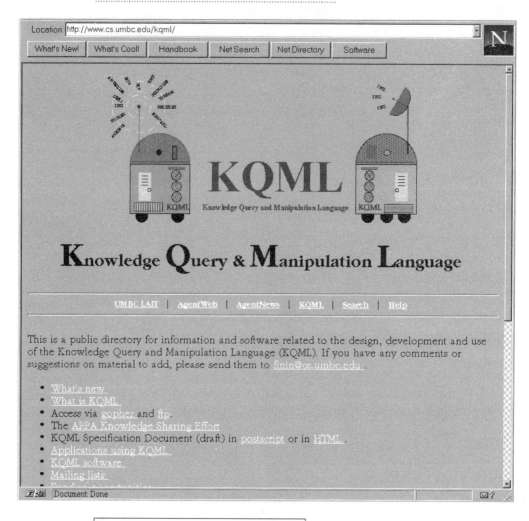

Location: http://www.cs.umbc.edu/kqml/

| What's New! | What's Cool! | Handbook | Net Search | Net Directory | Software |

Knowledge Query and Manipulation Language

**K**nowledge **Q**uery & **M**anipulation **L**anguage

UMBC LAIT | AgentWeb | AgentNews | KQML | Search | Help

This is a public directory for information and software related to the design, development and use of the Knowledge Query and Manipulation Language (KQML). If you have any comments or suggestions on material to add, please send them to finin@cs.umbc.edu .

- What's new
- What is KQML
- Access via gopher and ftp
- The ARPA Knowledge Sharing Effort
- KQML Specification Document (draft) in postscript or in HTML .
- Applications using KQML
- KQML software
- Mailing lists

Document: Done

## http://www.cs.umbc.edu/kqml

This useful Web site comprises a public directory for information, and software related to the design, development, and use of the Knowledge Query and Manipulation Language (KQML).

KQML is part of a larger project called the ARPA Knowledge Sharing Effort, which is aimed at developing techniques and methodology for building large-scale knowledge bases that are shareable and reusable. KQML itself is both a message format and a message-handling protocol to support run-time knowledge sharing among agents. KQML can be used as a language for an application program to interact with an intelligent system or for two or more intelligent systems to share knowledge in support of cooperative problem solving.

At this Web site you have free access to the KQML compiler software, test applications that use KQML, papers on KQML and related topics, information on KQML conferences and workshops, and the complete KQML draft specification document available in both PostScript and HTML formats.

# UNIVERSITY OF LONDON: QUEEN MARY AND WESTFIELD COLLEGE DISTRIBUTED ARTIFICIAL INTELLIGENCE RESEARCH UNIT

http://www.elec.qmw.ac.uk/dai/

Directed by Dr. Nick Jennings, the Distributed Artificial Intelligence (DAI) Research Unit at the University of London's Queen Mary and Westfield College has developed and applied DAI and agent-based techniques to real world problems in industrial and business domains. The unit has also worked on formalizing a number of key types of behavior that can be observed in multiagent systems. This research has used a variety of formal techniques (including modal and dynamic logics, and graph-based search) to develop detailed specifications of the processes of cooperative problem solving and of coordination in multiagent systems.

The DAI Web site includes the full texts of numerous papers by DAI unit members on such topics as social level characterizations of responsible agents, multiagent negotiation, shared ontologies for the integration of disparate agencies, negotiation between intelligent agents, coordinating agents for telecommunications applications, and information gathering agents.

# THE MAINE COOPERATIVE DISTRIBUTED PROBLEM-SOLVING RESEARCH GROUP

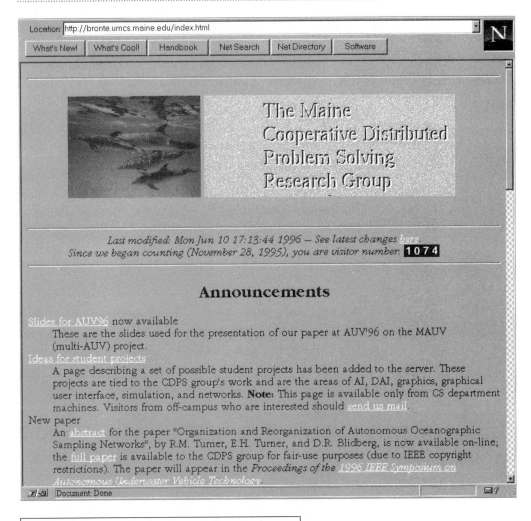

Location: http://bronte.umcs.maine.edu/index.html

| What's New! | What's Cool! | Handbook | Net Search | Net Directory | Software |

The Maine
Cooperative Distributed
Problem Solving
Research Group

*Last modified: Mon Jun 10 17:13:44 1996 — See latest changes here.*
*Since we began counting (November 28, 1995), you are visitor number:* **1074**

## Announcements

Slides for AUV96 now available
    These are the slides used for the presentation of our paper at AUV'96 on the MAUV
(multi-AUV) project.
Ideas for student projects
    A page describing a set of possible student projects has been added to the server. These
projects are tied to the CDPS group's work and are the areas of AI, DAI, graphics, graphical
user interface, simulation, and networks. **Note:** This page is available only from CS department
machines. Visitors from off-campus who are interested should send us mail.
New paper
    An abstract for the paper "Organization and Reorganization of Autonomous Oceanographic
Sampling Networks", by R.M. Turner, E.H. Turner, and D.R. Blidberg, is now available on-line;
the full paper is available to the CDPS group for fair-use purposes (due to IEEE copyright
restrictions). The paper will appear in the *Proceedings of the 1996 IEEE Symposium on
Autonomous Underwater Vehicle Technology.*

Document: Done

---

**http://bronte.umcs.maine.edu/index.html**

The Cooperative Distributed Problem-Solving (CDPS) Research
Group of the Department of Computer Science at the University of
Maine (in collaboration with the Marine Systems Engineering Labo-
ratory [MSEL] of Northeastern University) focuses on the problem of

getting autonomous, intelligent agents to cooperate with one another to solve complicated problems.

The primary domain of the CDPS Research Group is cooperative problem solving by groups of autonomous underwater vehicles (AUVs). Ongoing research projects involve intelligent AUV control, communication between autonomous agents, and coordination of problem solving by groups of autonomous agents.

The CDPS Research Group was founded in 1990 at the University of New Hampshire. In the fall of 1995, the group moved to the University of Maine when its founders, Roy and Elise Turner, began new appointments as assistant professors at the latter institution.

The CDPS Research Group Web site includes an extensive archive of group publications, most of them available in hypertext and/or PostScript formats.

# THE UNIVERSITY OF MASSACHUSETTS DISTRIBUTED ARTIFICIAL INTELLIGENCE LABORATORY

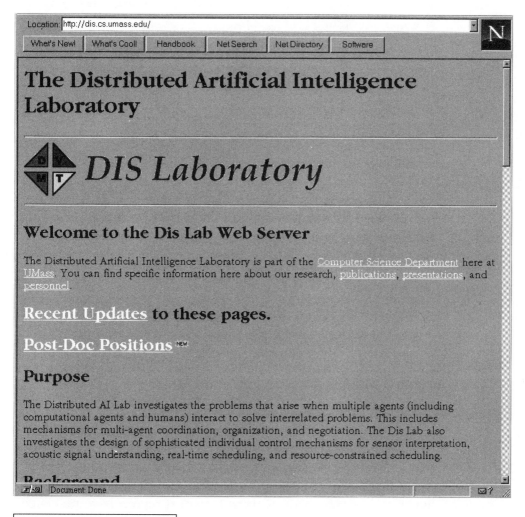

## The Distributed Artificial Intelligence Laboratory

### DIS Laboratory

### Welcome to the Dis Lab Web Server

The Distributed Artificial Intelligence Laboratory is part of the Computer Science Department here at UMass. You can find specific information here about our research, publications, presentations, and personnel.

### Recent Updates to these pages.

### Post-Doc Positions NEW

### Purpose

The Distributed AI Lab investigates the problems that arise when multiple agents (including computational agents and humans) interact to solve interrelated problems. This includes mechanisms for multi-agent coordination, organization, and negotiation. The Dis Lab also investigates the design of sophisticated individual control mechanisms for sensor interpretation, acoustic signal understanding, real-time scheduling, and resource-constrained scheduling.

Background

**http://dis.cs.umass.edu**

Led by Victor R. Lesser, the UMass Distributed AI Lab (DIS) investigates the problems that arise when multiple agents (including computational agents and humans) interact to solve interrelated problems. This includes mechanisms for multiagent coordination, organization, and negotiation. The DIS Lab also investigates the design of sophisticated individual control mechanisms for sensor interpretation, acoustic signal understanding, real-time scheduling, and resource-constrained scheduling.

Come to the DIS Lab Web site for information on ongoing research into automated contracting, coalition formation, control issues in parallel knowledge-based systems, cooperation among heterogeneous agents, cooperative information gathering, design-to-time real-time scheduling, and formal analysis of the FA/C Distributed Problem-Solving Paradigm. Here you will also find information on research related to generic agent architectures for real-time distributed situation assessment, generic coordination strategies for agents, integrating decision making with real-time scheduling, learning in multiagent systems, and negotiation among computationally bounded self-interested agents as well as among knowledge-based scheduling agents.

# THE UNIVERSITY OF MASSACHUSETTS LABORATORY FOR PERCEPTUAL ROBOTICS

http://piglet.cs.umass.edu:4321/lpr.html

The Laboratory for Perceptual Robotics (LPR) conducts research into dexterous manipulation, mobile robot navigation, geometric reasoning, assembly planning, and the application of learning theory to robotics. Come to this Web site for detailed reports on various experiments at the Lab. Many of these hypertext reports have multimedia features and include MPEG movies of various robotics lab demos.

# UNIVERSITY OF MICHIGAN DISTRIBUTED INTELLIGENT AGENTS GROUP

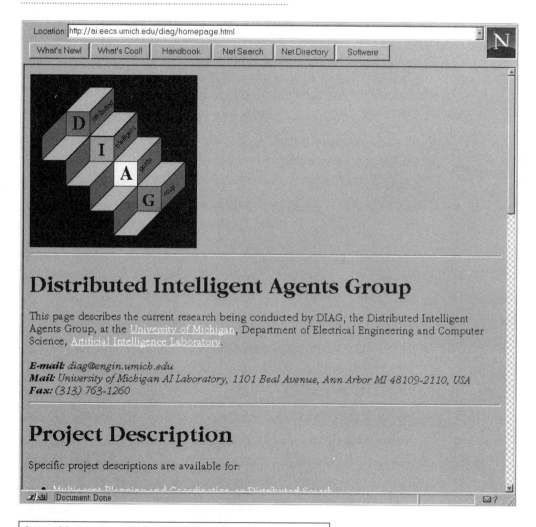

**http://ai.eecs.umich.edu/diag/homepage.html**

Headed by Edmund H. Durfee, the Distributed Intelligent Agents Group (DIAG) of the AI Laboratory at the University of Michigan has focused its research on several key areas:

- multiagent planning and coordination for a range of applications

- coordination through plan recognition

- organizational self-design

- recursive agent modeling

- and intelligent agent infrastructures for supporting collaboration.

Come to the DIAG Web site for information and progress reports on all these various research pathways, but also for a great article by Durfee that is a "must" read: "What Your Computer Really Needs to Know, You Learned in Kindergarten." Robert Fulgum gone cyber. Check it out.

# MIT ARTIFICIAL INTELLIGENCE LABORATORY

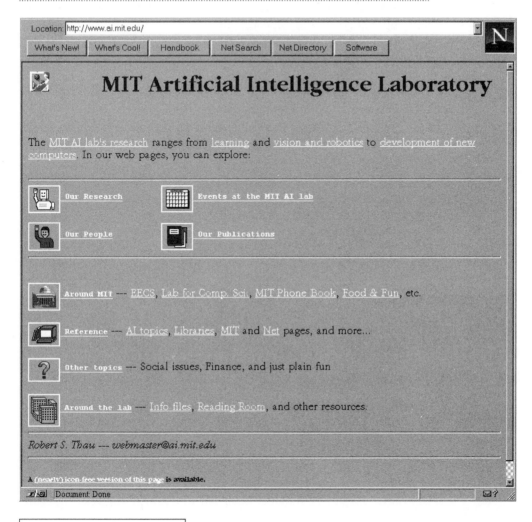

**http://www.ai.mit.edu/**

The MIT AI Lab's research ranges from learning and vision and robotics to development of new computers. In the Lab's Web pages you can explore details on MIT research into computer vision, mobile robotics, microbotics, haptic interfaces, vision- and touch-guided manipulation, robot hands, and more. Here you will also find information regarding the ongoing work of the Humanoid Robotics Project (The Cog Shop) at the MIT Lab and goings on within the "Zoo," which is a collection of people focused on understanding human intelligence and enriching the research centered in The Cog Shop.

Additionally, this Web site offers information on MIT Artificial Muscle Project, the MIT Trainable Modular Vision System, and the Center for Biological and Computational Learning, which provides a high-performance computing environment to foster the development of multidisciplinary approaches to the study of problems in the fields of intelligence and learning.

The site also provides in-depth information on:

❏ the Intelligent Information Infrastructure Project, which researches general systems for distributing and retrieving information and has led to the construction of automated tools for managing outbound and inbound communications flows whether via e-mail, distributed hypermedia, or other electronic media as well as interactive tools for wide-area communication, including approaches to natural language understanding;

❏ the InfoLab Project, which involves the design, development, and demonstration of software technology to help humans solve problems in *human terms*, interacting through the concentrated use of simple language, images, diagrams, and other modes of expression that are intrinsically meaningful to humans and engage intuitive human problem-solving skills.

# NEURALITY: THE NEURAL NETWORKS META-SITE

**http://www.icenet.it/icenet/neurality/home_uk.html**

Neurality represents the cyber crossroads for all the coolest and most useful information on neural networks to be found on-line, and includes current research topics relating neural applications.

Come here to check out:

❒ DamnNet—a neural net that knows how to design arch dams;

❒ NeuroTetris—download this great freeware neural network version of the legendary game;

❒ Neuro Fractals—neural networks that recognize and plot fractals;

❒ Cop-Net—a neural network that filters noisy and "bad" data.

Beyond this, Neurality provides a long list of links of value to developers and researchers, including pointers to the home pages of societies that include the International Neural Network Society (INNS), the European Neural Network Society (ENNS), the Japanese Neural Network Society (INNS), Foundation for Neural Networks (SNN), and IEEE Neural Network Council (NNC).

There is also a great set of links to demo neural network software and a stimulating HTML introduction to neural nets. The brief introduction to the fundamentals of neural nets includes an example involving baseball (a lot of fun). And the free downloadable software includes *Neuron* (very interesting software for simulating biological neurons), the *Discern* story-processing demo, which demonstrates a neural network at work, and a really strong simulated annealing demo.

Additional pointers take you to gateways to an extensive lists of neural network Web journals, the Astin Neural Network Lab list of global neural network research groups around the world, the official

Web page for the ALVINN (Autonomous Land Vehicle in a Neural Network) Project, and the Neural Archive at FUNet. And don't forget CERN's site documenting neural networks in high-energy physics, or the great site entitled *The Backpropagator's Online Reading List and Review* (packed with useful tricks for BackPropagation geeks!)

# ROBOTICS SITES ON THE WEB

University of Amsterdam Robotics and Neural Computing Page
**http://www.fwi.uva.nl/research/neural**

Cal Tech Robotics Home Page
**http://robby.caltech.edu**

University of California—Berkeley Human Engineering Lab
**http://www.me.berkeley.edu/hel/**

University of California—Berkeley Robotics and Intelligent Machines Lab
**http://robotics.eecs.berkeley.edu/**

University of California—Los Angeles Commotion Laboratory
**http://muster.cs.ucla.edu:8001/**

Cambridge University Speech, Vision, and Robotics Group
**http://svr-www.eng.cam.ac.uk/**

Carnegie-Mellon Robotics Institute
**http://www.ri.cmu.edu/**

Carnegie-Mellon Vision Page
**http://www.cs.cmu.edu/~cil/vision.html**

Columbia University Robotics Lab
**http://www.cs.columbia.edu/robotics/**

Cornell University Robotics and Vision Laboratory

**http://www.cs.cornell.edu/Info/Projects/csrv/csrv.html**

Indiana University Robotics Page

**http://www.cs.indiana.edu/robotics/robotics.html**

Johns Hopkins Robotics Lab

**http://caesar.me.jhu.edu/**

University of Manchester Robotics Lab

**http://www.cs.man.ac.uk/robotics**

McGill University Center for Intelligent Machines

**http://www.cim.mcgill.ca/**

New York University Center for Advanced Technology

**http://found.cs.nyu.edu**

Northwestern University Intelligent Perception and Action Lab

**http://tinkertoy.ils.nwu.edu/**

Oxford University Robotics Research Group

**http://www.robots.ox.ac.uk:5000/**

Robotics Internet Resources Page

**http://piglet.cs.umass.edu:4321/robotics.html**

Stanford Dextrous Manipulation Lab

**http://cdr.stanford.edu/html/Touch/touchpage.html**

University of Surrey Mechatronic Systems and Robotics Research Group

**http://robots.surrey.ac.uk/**

Yale Vision and Robotics Group

**http://www.cs.yale.edu/HTML/YALE/VISION/GroupPR.html**

# ARTIFICIAL LIFE

Artificial Life Resources
Brandeis University ALife Papers Archive
The Live Artificial Life Page
Macintosh Artificial Life Software
Marco's Maddening Artificial Life Page

George Mason University Genetic Algorithms Group
MIT Press Artificial Life Online
NanoTechnology Magazine
Zooland

# ARTIFICIAL LIFE RESOURCES

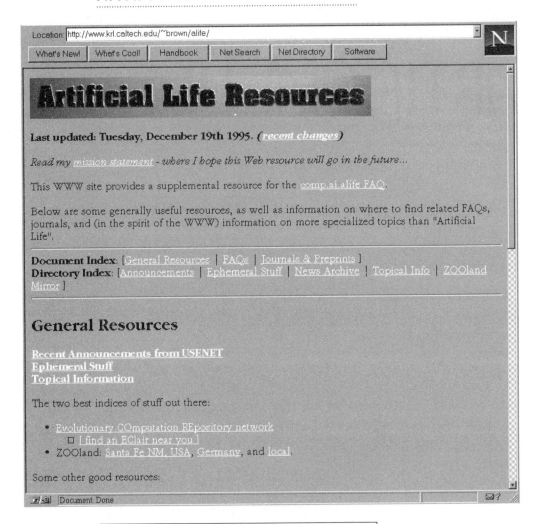

**http://www.krl.caltech.edu/~brown/alife**

Maintained by the indomitable Titus Brown at Caltech, this excellent list of links to resources is strongest in the archives it provides for major Artificial Life newsgroups, and also in two exceptional on-line bibliographies related to complex systems and ALife.

# BRANDEIS UNIVERSITY ALIFE PAPERS ARCHIVE

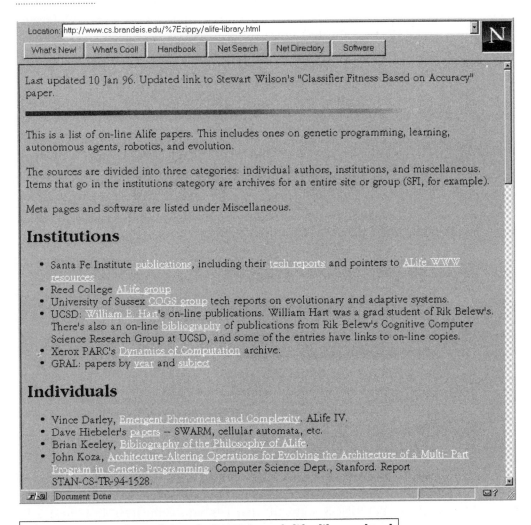

Location: http://www.cs.brandeis.edu/%7Ezippy/alife-library.html

| What's New! | What's Cool! | Handbook | Net Search | Net Directory | Software |

Last updated 10 Jan 96. Updated link to Stewart Wilson's "Classifier Fitness Based on Accuracy" paper.

This is a list of on-line Alife papers. This includes ones on genetic programming, learning, autonomous agents, robotics, and evolution.

The sources are divided into three categories: individual authors, institutions, and miscellaneous. Items that go in the institutions category are archives for an entire site or group (SFI, for example).

Meta pages and software are listed under Miscellaneous.

## Institutions

- Santa Fe Institute publications, including their tech reports and pointers to ALife WWW resources
- Reed College ALife group
- University of Sussex COGS group tech reports on evolutionary and adaptive systems.
- UCSD: William E. Hart's on-line publications. William Hart was a grad student of Rik Belew's. There's also an on-line bibliography of publications from Rik Belew's Cognitive Computer Science Research Group at UCSD, and some of the entries have links to on-line copies.
- Xerox PARC's Dynamics of Computation archive.
- GRAL: papers by year and subject

## Individuals

- Vince Darley, Emergent Phenomena and Complexity, ALife IV.
- Dave Hiebeler's papers -- SWARM, cellular automata, etc.
- Brian Keeley, Bibliography of the Philosophy of ALife.
- John Koza, Architecture-Altering Operations for Evolving the Architecture of a Multi-Part Program in Genetic Programming, Computer Science Dept., Stanford. Report STAN-CS-TR-94-1528.

Document Done

**http://www.cs.brandeis.edu/%7Ezippy/alife_library.html**

Patrick Tufts created and maintains this wonderful set of pointers to virtually all of the currently available on-line papers concerning genetic programming, learning, autonomous agents, robotics, and evolution.

Tufts gives you links to great institutional archives that include the papers of the Reed College ALife Group, UCSD's archive of William E. Hart's seminal research as well as a great bibliography of publications from Rik Belew's Cognitive Computer Science Research Group at UCSD, and the full contents of Xerox PARC's Dynamic of Computation archive as well as the genetic programming archive at the University of Texas.

More importantly, Tufts provides links to individual archives of papers by many of the leading researchers in the field of ALife. Here you will find HTML and PostScript editions of Vince Darley's papers on emergent phenomena and complexity, Dave Heibeler's papers on SWARM and cellular automata, Brian Keeley's indispensable *Bibliography of the Philosophy of ALife*, and John Koza's major paper on architecture-altering operations for evolving the architecture of a multipart program in genetic programming.

Additional links take you to Maja Mataric's papers on interaction and intelligent behavior, reward functions for accelerated learning, kin recognition, and behavior-based systems; Filippo Menczer's papers on the use of ALife models applied to the construction of tools for natural (biological) and artificial (informational) ecologies; Craig Reynolds's research on the evolution of corridor following behavior in a noisy world; Justinian Rosca's and Dana Ballard's groundbreaking consideration of genetic programming with adaptive representations; and Luc Steels's extensive papers on such topics as building agents with autonomous behavior systems.

Additional pointers deliver Karl Sims's paper and companion MPEG film on evolving virtual creatures, first prepared for SIGGraph '94; downloadable software for Polyworld and other ALife applications; and the POPBUGS software that lets you simulate robots in a 2-D world.

# LIVE ARTIFICIAL LIFE PAGE

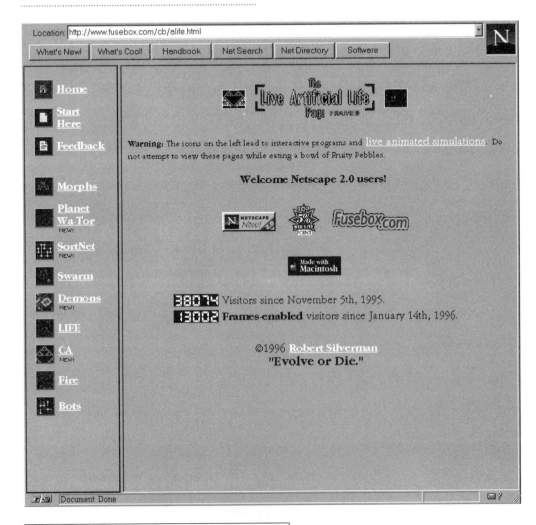

http://www.fusebox.com/cb/alife.html

Start by heeding the warning that meets you on the top screen of this supercool Web site: "The following links lead to interactive programs and live animated simulations. Do not attempt to view these pages while eating a bowl of Fruity Pebbles."

Quite simply, you have some awesome ALife-related things to download and/or play with at this techno-wonderland:

- ❐ Morphs—An adaptation of Richard Dawkins's Biomorphs as seen in *The Blind Watchmaker*. Evolve and grow your own community of pet morphs; then pick your favorite morph and post it online for others to see!

- ❐ PlanetWa-Tor—A.K. Dewdney's ecology toy from *The Armchair Universe* lets you set the controls and then watch the sharks drive fish to the brink of extinction.

- ❐ SortNet—Use a simple genetic algorithm to evolve minimal sorting networks in this "lite" version of Daniel Hillis's Ramps software.

- ❐ Swarm—Watch flocking behavior spontaneously emerge in this adaptation of Craig Reynolds's Boids.

- ❐ Demons—Try to get the demons to emerge in David Griffeath's cyclic space automata.

- ❐ Life—Yes! Conway's Life! The program that started it all is now on the Web!

- ❐ CA—You are at the controls of Stephen Wolfram's 1-Dimensional Cellular Automata!

- ❐ Bots—Robert "Evolve-or-Die" Silverman's fantastic creation wherein you can watch some critters exhibiting behavior based on their local environment and their genes.

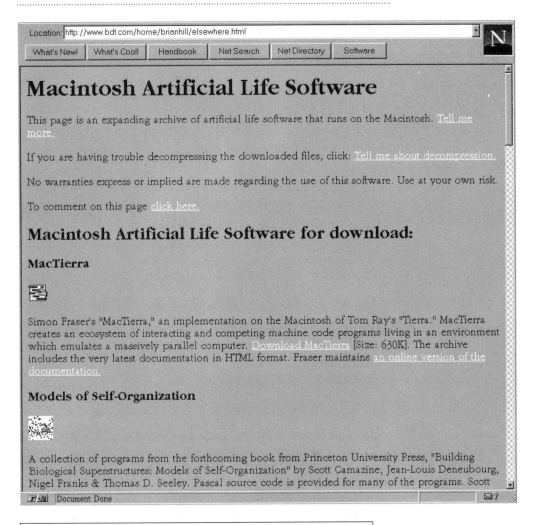

**http://www.bdt.com/home/brianhill/elsewhere.html**

Here is some great ALife software for all you Mac-heads to download and have fun with ... I mean, it is software for you to study and use as tools in your ongoing ALife research. (That's the way you should word it for the grant proposal.)

Just check out these items:

Simon Fraser's MacTierra is a Mac implementation of Tom Ray's classic Tierra. MacTierra creates an ecosystem of interacting and competing machine code programs living in an environment that emulates a massively parallel computer.

Dave Johnson's Evolve-o-Matic is a Macintosh environment in which each critter has a set of "genes" that determine its movement patterns and its breeding behavior. Some of a critter's genes control its response to a sun that moves around the world. The critter can be attracted or repelled, excited or calmed, by light. And each critter has a "soft" or "sensitive" side which it may try to keep toward or away from the light. Combinations of these factors lead to interesting behaviors: orbiting, sun humping (don't ask), sling shooting, etc. In addition to downloading Evolve-o-Matic be sure to download the neat file of saved worlds that shows highly evolved creatures that can emerge from this program after thousands of generations.

Kevin Coble's Neoterics is a program in which creatures evolve neural nets to guide them in gathering food and mating. (Note that this program will not run on older Macs such as the Plus or SE. Also, if you have a PowerMac 6100 or higher, be sure to download PowerMac Neoterics, which will run faster on your platform.)

Keith Wiley's Bugs is a Macintosh environment of plants, rocks and (of course!) hungry bugs. Menus allow you to observe and manipulate the distribution of genes among the evolving bug population.

Ryan Koopman's Vivarium allows you, the user, to specify the general strategy of an organism (for example, grazer, predator, cannibal, or some combination). The organism must then evolve rules governing its response to the world. What knowledge, for example, would cannibalistic slugs take away from their first encounter with predator spiders?

Alex Vulliamy's Flies is a simple demonstration for Macintosh of a flocking theory. Each fly in the environment tries to maintain itself near its two closest neighbors, leading to an emergent behavior in which the flock seems to move with a single mind. (This is even more hypnotic than a lava lamp.)

Brian Hill's TargetSeekers is an environment in which evolving algorithms for hitting the center of a target compete with evolving target defense patterns.

Tresvits 1.0, Alexander Kasprzyk's three-dimensional Life program, allows the user to set rules under which cells live, die, and are reborn, and to create custom cell patterns.

Thai Truong's Mac Cellular Automata is a demo of cellular automata following one-dimensional rules. This program shows that simple local rules, repeated often enough, can result in complex and amazing global patterns. Fred J. Condo, Jr.'s Hodgepodge(Life) demos cellular automata following two-dimensional rules. This program uses a hypothetical model for the spread of disease. Interesting patterns emerge as cells infect each other, die, and are reborn. And lastly, Reggie McLeod's Easy Life 2.0 is the Mac version of the most widely known two-dimensional cellular automata, LIFE. This program shows several forms of cyclical, recurring, and coherent moving forms that are possible under the simple set of LIFE rules.

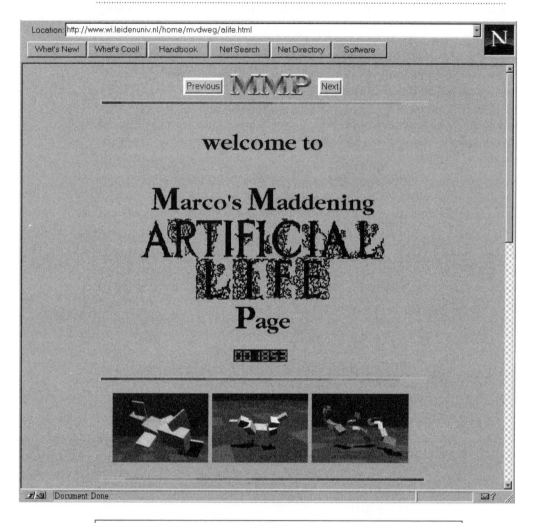

**http://www.wi.leidenuniv.nl/home/mvdweg/alife.html**

Marco provides a great set of links that includes connections to the Web pages for the Artificial Life Group at Iowa State, the Avida Group at the California Institute of Technology, the Autonomous Agents Research Group at Case Western Reserve, and ALife archives at the University of Pennsylvania. You also get access to PC software for

C.O.P.P. with Biotopia 2.0 (a Darwinistic physical ecosystem by Anthony Liekens), GeNeura (the evolution game), and an MPEG version of Karl Sims's film Virtual Creatures.

# GEORGE MASON UNIVERSITY GENETIC ALGORITHMS GROUP

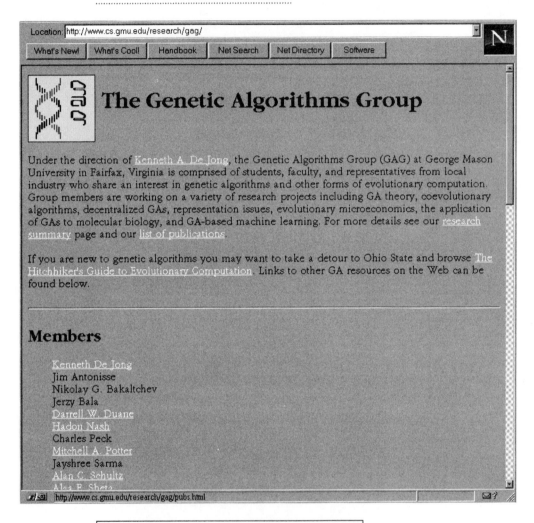

Group members are working on a variety of research projects including GA theory, coevolutionary algorithms, decentralized GAs, representation issues, evolutionary microeconomics, the applications of GAs to molecular biology, and GA-based machine learning. Here at the GAG Web site you will find research summaries, a list of publications, and a cool link to the Ohio State server where you can browse through something absolutely great: Jörg Heitkötter's *The Hitchhiker's Guide to Evolutionary Computation*. Go for it.

# MIT PRESS ARTIFICIAL LIFE ON-LINE

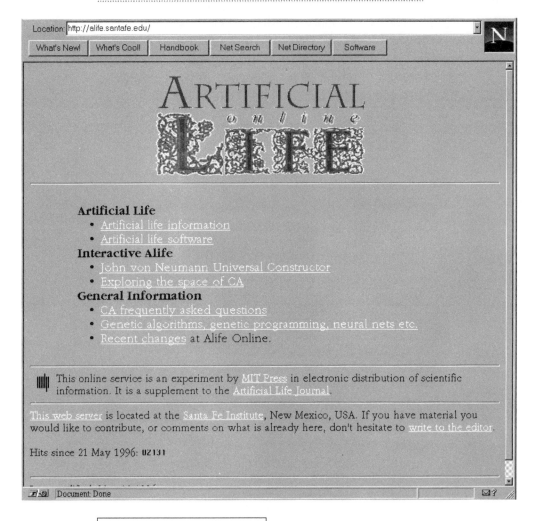

http://alife.santafe.edu/

This site on the Web offers far fewer bells, whistles, and toys than other ALife sites, but one thing it has going for it is a splendid, regularly updated listing of upcoming ALife events, symposia, and calls for papers. There is no better spot in cyberspace to find so extensive a listing of ALife conferences, meetings, and conventions. Here you will also find abstracts and subscription information for MIT's three ALife-related journals, *Adaptive Behavior*, *Evolutionary Computation*, and *ALife Digest*.

# NANOTECHNOLOGY MAGAZINE

http://planet-hawaii.com/nanozine/

*NanoTechnology* Magazine is a fantastic window into the emerging technology related to the atomic manipulation matter. Via this magazine and this Web site you can follow monthly discoveries involved in the evolution of a technology that is likely to dominate the twenty-first century.

# ZOOLAND

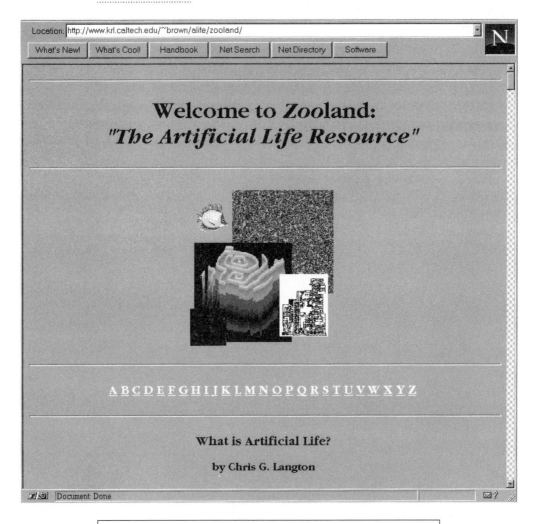

http://www.krl.caltech.edu/~brown/alife/zooland/

There is simply no other set of ALife resources on the Web to match that presented at Zooland.

Zooland gives you links to:

❐ Afarm software for UNIX—Evolution in your pocket!

❑ ARVA UNIX software for simulating multiagent systems

❑ À la A.K. Dewdney, Joshua Smith's Biomorphs software for UNIX and PC-Windows, Core Wars software for UNIX and PC-Windows, and Jörg Heitkötter's Hodgepodge Machine for UNIX

❑ Andrew Wuensche's Discrete Dynamics Lab software for PC-Windows, with which to research the dynamics of finite binary networks

❑ Aquarium—Ric Colasanti's PC-Windows fish-tank simulator based on Craig Reynolds's ALife program Boids

❑ Avida—a 2D version of Tierra for UNIX by Chris Adami

❑ Craig Reynold's original Boids for UNIX as well as Jürgen Schmitz's Boids for PC-Windows

❑ Martyn Amos's BugWorld for UNIX

❑ Rudy Rucker's great Calife, a 1D cellular automata simulator for PC-Windows

❑ Cellism, a cellular automata simulator written for UNIX by David Heibeler and Chris Langton

❑ Echo, an ecological simulation system for UNIX by Terry Jones and John Holland

❑ JVN, an implementation of the John von Neumann Universal Constructor created for UNIX and PC-Windows by R. Nobili and U. Pesavento.

Additionally you get access to LifeLab Mac software by Andrew Trevorrow, David Bell's Life Search software for UNIX, the MbitiWorld PC-Windows program for creating neural-net carnivorous/herbivorous agents that evolve a bit and then eat each other, the Macrophylon PC-Windows software developed to investigate the patterns and dynamics involved in the building of Darwinian evolutionary trees, and more:

- ❏ Mitchell Resnik's Starlogo for Macintosh, a basic complex system implementation written in Logo

- ❏ Marc de Groot's Primordial Soup software for UNIX—an artificial life system that spontaneously generates self-reproducing organisms from a sterile soup

- ❏ Tierra—PC Windows and UNIX versions of Tom Ray's original system for studying ecological and evolutionary dynamics

- ❏ WinCA—a fast cellular automata simulator written by Bob Fisch and David Griffeath for PC-Windows

- ❏ Xca—Chris Langton's self-replicating cellular automaton for UNIX.

Finally, users of PC-Windows must be sure to treat themselves to John Harper's WinLife program, a Windows version of John Conway's "Game of Life." With WinLife you can create patterns and watch them grow according to the rules of the game. Thanks to the program's powerful editing facilities, you can make changes to complex patterns without reentering them. And you can save them to disk so that you can replay them later.

Add to all this great software many PostScript and HTML editions of papers by Howard Gutowitz, Gene Levinson, Nick Turner, David Bell, and other leading ALife researchers, and you start to realize that Zooland is absolutely the best source for ALife information and tools on the Web.

# ASTRONOMY

All-Sky Low Energy Gamma Ray Observatory (ALLEGRO)
American Astronomical Society
The ATM (Amateur Telescope Makers) Page
APS Catalog of POSS I
Armagh Observatory
**The Astronomer** Online
The Astronomy Cafe
Astronomy Image Library at UMASS
**Astronomy** Magazine
AstroWeb
Catalog of Infrared Observations
Comet C/1996 B2 Hyakutake
Comet Hale-Bopp Home Page
Digital Sky Survey
The European Space Agency
Galactic Sky Charts
Project Galileo: Bringing Jupiter to Earth
The Gemini 8-Meter Telescopes Project

HST's (Hubble Space Telescope's) Greatest Hits 1990–1995
Hypertext Astronomy Textbooks
Infrared Space Observatory
Limiting Magnitude
The MACHO Project
Multiwavelength Atlas of Galaxies
NASA
The Nine Planets: A Multimedia Tour
Parkes Multibeam Survey
SERENDIP: Search for Extraterrestrial Radio Emissions from
        Nearby Developed Intelligent Populations
SETI Institute Home Page
Sky Online
SkyView
Sloan Digital Sky Survey Science Archive
**Star Facts**: An Electronic Journal About the Universe
The Web Nebulae
Wisconsin H-Alpha Mapper (WHAM)

# ALL-SKY LOW ENERGY GAMMA RAY OBSERVATORY (ALLEGRO)

http://www.astro.nmu.edu/astro/allegro

ALLEGRO is a proposed MidEx class instrument providing all-sky monitoring of low-energy gamma rays at unprecedented sensitivity. Unlike previous hard x-ray experiments, there is no time averaging, data selection, or triggering on board. ALLEGRO transmits all events, time-tagged to 1/8th ms and with full energy information. This produces a database of uniformly high resolution in both energy and time, permitting non-triggered, unbiased detection of transient and pulsed events. Access that database here.

# AMERICAN ASTRONOMICAL SOCIETY

http://www.aas.org

The American Astronomical Society (AAS) is the major professional organization in North America for astronomers and other scientists and individuals interested in astronomy. Come to this Web site for information on the society's executive council, committees, education and grants programs, and publications. Come here most especially for the splendid electronic edition of the society's *Astrophysical Journal*.

# APS CATALOG OF POSS 1

http://isis.spa.umn.edu/homepage.aps.html

This Web site contains millions of entries for stars and galaxies, and the corresponding image database contains their pixels and more. The data behind the object catalog and image database are generated from digitized Palomar Observatory Sky Survey (POSS) plates. The object catalog entries include calibrated magnitudes in two colors, positions to 0.2 arcsecants, confidence measures on neural network image classifications, colors, and various other useful parameters.

# ARMAGH OBSERVATORY

http://starm.arm.ac.uk/

Founded in 1790, Armagh Observatory is located in North Ireland and is one of two sites that comprise the Northern Ireland node of the International Starlink network of observatories.

# THE ASTRONOMER ON-LINE

http://www.demon.co.uk/astronomer

This is the home page for The Astronomer Group (TA for short) based in the United Kingdom. TA has produced the magazine *The Astronomer* for advanced astronomers since 1964. The aim of *The Astronomer* is to publish all observations of astronomical interest as soon as possible after they are made.

Come to this Web site for information on subscribing to the paper edition of *The Astronomer*, but also for much more. Here you can sign up to receive electronic circular (distributed via e-mail) designed to keep you abreast of the latest-breaking celestial events. Here you can also access a number of Web pages designed by the TA staff and packed with information, images, and even on-line films related to comets and asteroids, novas and supernovas, variables (including observations of HT Cas in outburst), and specific planets.

# ASTRONOMY CAFE

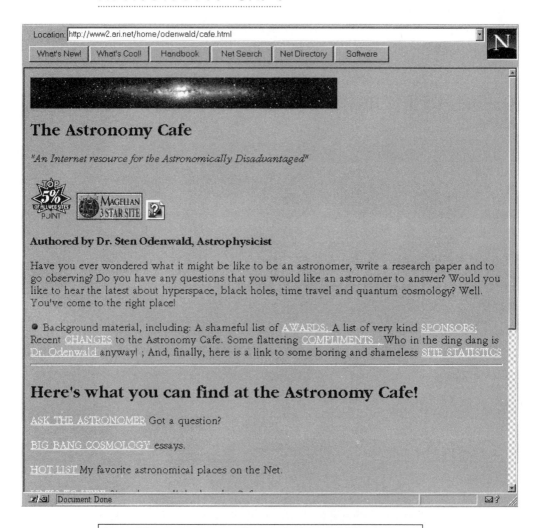

http://www2.ari.net/home/odenwald/cafe.html

Created and maintained by the astrophysicist Dr. Sten Odenwald, the Astronomy Cafe bills itself as a resource for "the astronomically disadvantaged." In other words, it is for people who are not astronomers but who have an interest in astronomy.

Got a specific question about astronomy? Post a note and Dr. Odenwald will address your query. Thinking about a career in astronomy? Access Odenwald's file in which he explains what is involved educationally, scientifically, and financially. What is the anatomy of a well-researched and well-written research paper? Odenwald will fill you in. Or you can answer a few questions designed to help you find out if you have "what it takes" to be an astronomer. There are also links to Odenwald's favorite astronomical places on the Net, and even sound files featuring music with an astronomical bent.

All in all, this is quite a fun spot on the Web for the astronomically challenged and the astronomically disadvantaged.

# ASTRONOMY IMAGE LIBRARY AT UMASS

**http://donald.phast.umass.edu/gs/wizimlib.html**

Here is your door to an extensive collection of outstanding astronomical resources that include:

❐ a hyperlinked tour of the Lynds 1641 molecular cloud near Orion

❐ a library of molecular line maps for star-formation regions

❐ a three-color infrared image of the Ophiuchus region, and another of the central section of the Serpens Star Forming region

❐ an infrared cluster associated with the NGC2024 nebula

❐ and a stunning image of Jupiter captured with the NICMASS camera.

# ASTRONOMY MAGAZINE

http://www.kalmbach.com/astro/astronomy.html

*Astronomy* Magazine is the world's largest English-language magazine for astronomy hobbyists. Produced by Kalmbach Publishing Company, *Astronomy* is read by over 300,000 people each month.

The *Astronomy* Magazine Web site provides a hypertext edition of one article from the current monthly edition of the magazine plus a hypertext edition of the current monthly "Sky Almanac" column, giving you details on celestial events to watch for. You also get complete TOCs for this month's and next month's magazine, an up-to-date listing of meetings and events, an on-line guide to scopes and accessories for the backyard astronomer, and reviews of astronomical software, books, and other merchandise.

Are you a writer or photographer? Go on-line to access a hypertext guide for submitting manuscripts and photos. Here you will also find an electronic listing of jobs opportunities at *Astronomy*.

# ASTROWEB

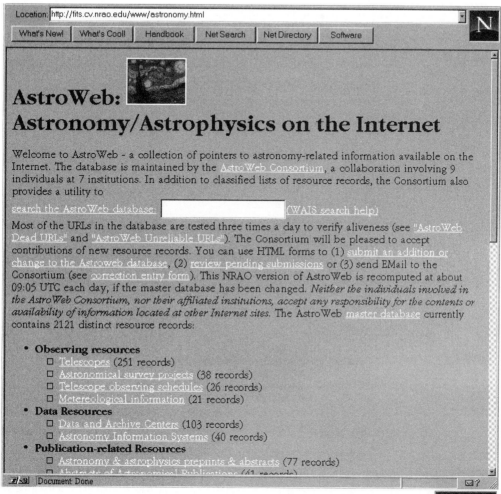

**AstroWeb:**
**Astronomy/Astrophysics on the Internet**

Welcome to AstroWeb - a collection of pointers to astronomy-related information available on the Internet. The database is maintained by the AstroWeb Consortium, a collaboration involving 9 individuals at 7 institutions. In addition to classified lists of resource records, the Consortium also provides a utility to

search the AstroWeb database: [                    ] (WAIS search help)

Most of the URLs in the database are tested three times a day to verify aliveness (see "AstroWeb Dead URLs" and "AstroWeb Unreliable URLs"). The Consortium will be pleased to accept contributions of new resource records. You can use HTML forms to (1) submit an addition or change to the Astroweb database, (2) review pending submissions or (3) send EMail to the Consortium (see correction entry form). This NRAO version of AstroWeb is recomputed at about 09:05 UTC each day, if the master database has been changed. *Neither the individuals involved in the AstroWeb Consortium, nor their affiliated institutions, accept any responsibility for the contents or availability of information located at other Internet sites.* The AstroWeb master database currently contains 2121 distinct resource records:

- **Observing resources**
  - □ Telescopes (251 records)
  - □ Astronomical survey projects (38 records)
  - □ Telescope observing schedules (26 records)
  - □ Metereological information (21 records)
- **Data Resources**
  - □ Data and Archive Centers (103 records)
  - □ Astronomy Information Systems (40 records)
- **Publication-related Resources**
  - □ Astronomy & astrophysics preprints & abstracts (77 records)
  - □ Abstracts of Astronomical Publications (41 records)

Document: Done

---

**http://fits.cv.nrao.edu/www/astronomy.html**

AstroWeb is the best single collection of pointers to astronomy-related information on the Internet. The database is maintained by the AstroWeb Consortium, a collaboration involving nine individuals at seven institutions. This fully searchable database includes no less than 1,809 distinct resource records including:

❐ observing resources—telescopes (237 records), astronomical sur-vey projects (31 records), telescope observing schedules (25 records), and meteorological information (19 records);

❐ data resources—data and archive centers (102 records) and astronomy information systems (42 records);

❐ publication-related resources—astronomy and astrophysics pre-prints and abstracts (76 records), abstracts of astronomical publications (39 records), full texts of astronomical publications (52 records), astronomical bibliographical services (28 records), and astronomy-related libraries (22 records);

❐ people-related resources—personal Web pages of astronomers and astrophysicists (486 records), astronomy job listings (4 records), conferences and meetings (43 records), astronomy newsgroups (21 records), and astronomy mailing lists (23 records);

❐ organizations—astronomy departments (318 records), astronom-ical societies (53 records), and space agencies (42 records);

❐ software resources—astronomy software servers (119 records) and downloadable document preparation tools such as TeX (6 records);

❐ specific research areas—radio astronomy (76 records), optical astronomy (190 records), high-energy astronomy (30 records), space astronomy (135 records), solar astronomy (45 records), planetary astronomy (15 records), and the history of astronomy (8 records);

❐ more than 100 records of Web files containing astronomical images.

# ATM (AMATEUR TELESCOPE MAKERS) PAGE

http://www.tiac.net/users/atm/

The ATM Page is intended as a resource for both beginning and advanced makers of home-grown telescopes. The Web document contains a complete annotated hypertext bibliography of telescope-making as well as information on mountings and optics, a library of telescope designs, and a gallery of ATM scopes. For even more information, check out the Web page for *Amateur Telescope Making Magazine*, http://webspace.com/markv/telescopes.shtml.

# CATALOG OF INFRARED OBSERVATIONS

http://iuewww.gsfc.nasa.gov/cio/cio_homepage.html

The Catalog of Infrared Observations is a database of over 200,000 published infrared observations of more than 10,000 individual astronomical sources over the wavelength range from 1 to 1000 microns. Items in the catalog are available for downloading via ftp (File Transfer Protocol).

# COMET C/1996 B2 HYAKUTAKE

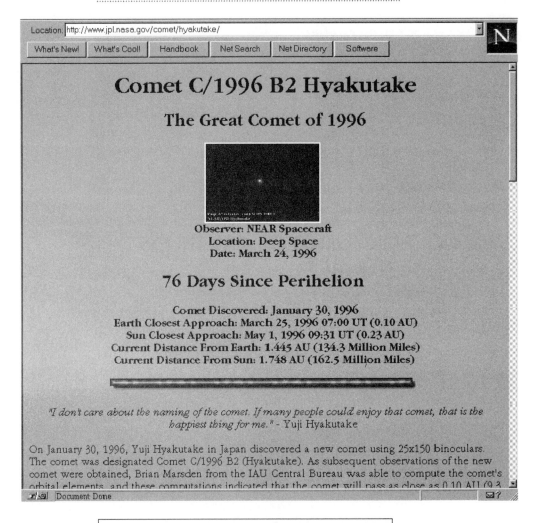

## Comet C/1996 B2 Hyakutake

### The Great Comet of 1996

Observer: NEAR Spacecraft
Location: Deep Space
Date: March 24, 1996

### 76 Days Since Perihelion

Comet Discovered: January 30, 1996
Earth Closest Approach: March 25, 1996 07:00 UT (0.10 AU)
Sun Closest Approach: May 1, 1996 09:31 UT (0.23 AU)
Current Distance From Earth: 1.445 AU (134.3 Million Miles)
Current Distance From Sun: 1.748 AU (162.5 Million Miles)

*"I don't care about the naming of the comet. If many people could enjoy that comet, that is the happiest thing for me."* - Yuji Hyakutake

On January 30, 1996, Yuji Hyakutake in Japan discovered a new comet using 25x150 binoculars. The comet was designated Comet C/1996 B2 (Hyakutake). As subsequent observations of the new comet were obtained, Brian Marsden from the IAU Central Bureau was able to compute the comet's orbital elements, and these computations indicated that the comet will pass as close as 0.10 AU (9.3

Document Done

---

**http://www.jpl.nasa.gov/comet/hyakutake**

On January 30, 1996, Yuji Hyakutake of Japan discovered a new comet using 25 × 150 binoculars. The comet was designated Comet C/1996 B2 (Hyakutake). As subsequent observations of the new comet were obtained, Brian Marsden of the International Astronomical Union (IAU) Central Bureau was able to compute the comet's oribital elements, and these computations indicated that the comet would pass as close as 0.11 AU (9.3 million miles) to Earth in March 1996. Come to this great Web site for an enormous collection of Comet Hyakutake images, observations, and data files.

# COMET HALE-BOPP HOME PAGE

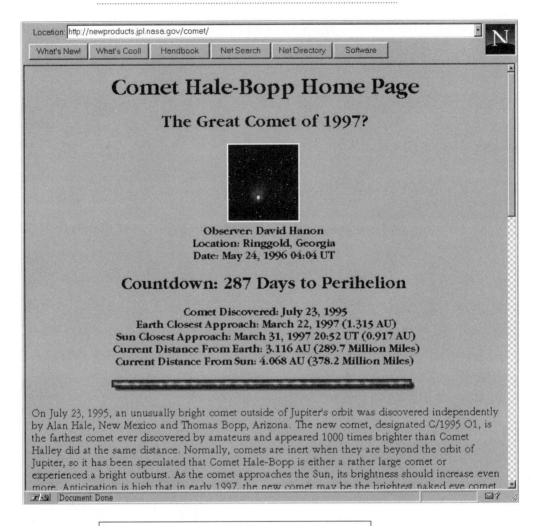

What's New! | What's Cool! | Handbook | Net Search | Net Directory | Software

## Comet Hale-Bopp Home Page

### The Great Comet of 1997?

Observer: David Hanon
Location: Ringgold, Georgia
Date: May 24, 1996 04:04 UT

## Countdown: 287 Days to Perihelion

Comet Discovered: July 23, 1995
Earth Closest Approach: March 22, 1997 (1.315 AU)
Sun Closest Approach: March 31, 1997 20:52 UT (0.917 AU)
Current Distance From Earth: 3.116 AU (289.7 Million Miles)
Current Distance From Sun: 4.068 AU (378.2 Million Miles)

On July 23, 1995, an unusually bright comet outside of Jupiter's orbit was discovered independently by Alan Hale, New Mexico and Thomas Bopp, Arizona. The new comet, designated C/1995 O1, is the farthest comet ever discovered by amateurs and appeared 1000 times brighter than Comet Halley did at the same distance. Normally, comets are inert when they are beyond the orbit of Jupiter, so it has been speculated that Comet Hale-Bopp is either a rather large comet or experienced a bright outburst. As the comet approaches the Sun, its brightness should increase even more. Anticipation is high that in early 1997, the new comet may be the brightest naked eye comet

Document: Done

**http://newproducts.jpl.nasa.gov/comet/**

On July 23, 1995, an unusually bright comet outside of Jupiter's orbit was discovered independently by Alan Hale of New Mexico and Thomas Bopp of Arizona. The new comet, designated C/1995 O1, is the farthest comet ever discovered by amateurs and appeared 1000 times brighter than Comet Halley did at the same distance. Normally, comets are inert when they are beyond the orbit of Jupiter, so it has been speculated that Comet Hale-Bopp is either a rather large comet or experienced a bright outburst. As the comet approaches the sun, its brightness should increase even more. Anticipation is high that in early 1997 the new comet may be the brightest naked eye comet since Comet West in 1976. Come to this home page for a complete, constantly updated gallery of images of Comet Hale-Bopp generated by the Hubble Space Telescope, the Teide Observatory, the Apache Point Observatory, and from other sources. As of this writing, there are no less than 133 images archived at this site. Come here also for electronic reprints of articles about the comet from various journals.

# DIGITAL SKY SURVEY

http://stdatu.stsci.edu/dss/

The Digital Sky Survey consists of the 1950/55 Palomar Observatory Sky Survey red plates for the northern sky and the SERC Southern Sky Survey (including the SERC J Equatorial Extension and some short V-band plates at low galactic latitude). And the results of the entire project are available here on-line. A convenient name resolver is available to help you find the coordinates of the object you are interested in. This site is a tremendous resource.

# EUROPEAN SPACE AGENCY

http://www.esrin.esa.it/

The European Space Agency promotes cooperation among European countries in space research and technology. Complete details on the organization—including its history, charter, and long-term projects— may be found at this Web page.

# GALACTIC SKY CHARTS

http://www.calweb.com/~mcharvey/

Who could ask for more? Use a clickable map of the world to access sky charts to help you plan this evening's viewing. You will get a chart that will be accurate for you at 8:00 P.M. local time, displaying a view from 40° latitude for your hemisphere. The charts can be viewed on any computer linked to the World Wide Web, but were created using Carina Software's Voyager II software for the Macintosh.

# THE GEMINI 8-METER TELESCOPES PROJECT

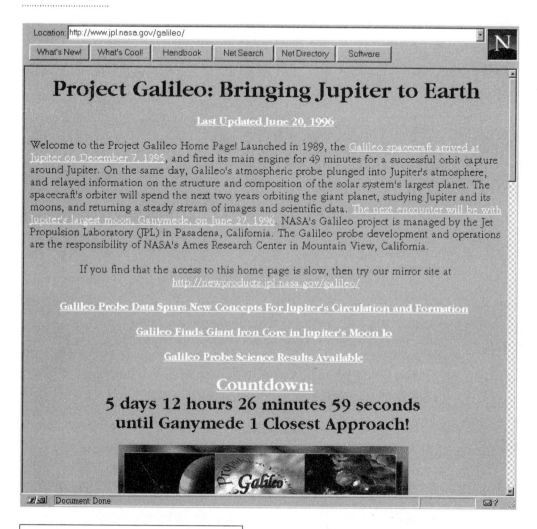

http://www.jpl.nasa.gov/galileo

The Gemini Project consists of two high-performance, 8-meter aperture optical/infrared telescopes in development as part of an international partnership between the governments of the United States, the United Kingdom, Canada, Chile, Argentina, and Brazil. The two telescopes are to be positioned in the Northern Hemisphere at Mauna Kea, Hawaii and in the Southern Hemisphere at Cerro Pachon, Chile. The scopes will be up and running by the year 2000. The goal is to exploit the best natural observing conditions in the world so that scientists from the partner countries can initiate a broad range of astronomical research programs that were heretofore impossible. Visit the Gemini Web site for many more details on the project.

# HST'S (HUBBLE SPACE TELESCOPE'S) GREATEST HITS 1990 --1995

http://www.stsci.edu/pubinfo/BestOfHST95.html

Access a splendid photo gallery of the universe generated by the Hubble Space Telescope including:

❏ Supernova 1987A: Halo for a Vanished Star—An eerie, nearly mirror-image pair of red luminescent gas "hula-hoops" frame the expanding debris of a star seen as a supernova explosion in 1987.

❏ The Orion Nebula: Stellar Birthplace—an immense wall of glowing gases forms a colorful backdrop to dozens of newborn stars, many of which have dust disks (as revealed by Hubble) that might be embryonic solar systems.

❏ The Cartwheel Galaxy: Result of a Bull's-Eye Collision—a spectacular head-on collision between a spiral galaxy and a smaller intruder sends out a ripple of energy that triggers a firestorm of new starbirth, forming a dazzling ring-like structure.

❏ Comet P/Shoemaker-Levy 9 Bombards Jupiter—Hubble followed unexpected and dramatic changes in Jupiter's atmosphere caused by collisions with comet fragments. The titanic blasts left Jupiter with a temporarily "bruised" appearance, caused by black debris that was tossed high above the giant planet's cloud tops.

- Storm on Saturn—a rare storm, large enough to swallow Earth, appears near Saturn's equator. High altitude winds give the storm a distinctive arrowhead shape.

- Ring Around a Suspected Black Hole in Galaxy NGC 4261—the gravitational pull of a suspected black hole forms a frisbee-like disk of cool gas, at the core of an energetic galaxy. Note that subsequent Hubble observations of yet another active galaxy (M87) confirmed the reality of monstrous black holes: gravitational "sink holes" that trap everything, even light.

- Planetary Nebula NGC 6543: Gaseous Cocoon Around a Dying Star—mysterious stellar fireworks create expanding gas shells and blowtorch-like jets that form a spectacularly intricate and symmetrical structure. The nebula is a fossil record of the late stages of the star's evolution.

- Cygnus Loop: Blast Wave from a Stellar Time Bomb—high-speed gas from a supernova explosion slams into dark cooler clouds of interstellar material. Shocked and heated by this tidal wave of energy, the clouds glow in bright, neon-like colors.

# HYPERTEXT ASTRONOMY TEXTBOOKS

http://zebu.uoregon.edu/text.html

Here the Electronic Universe Project provides you with several excellent hypertext tutorials on properties of galaxies, planetary motion, the inverse square law, the interstellar medium, stellar evolution, mucleosynthesis in stars, and the evolution of star clusters. These are multimedia hypertexts that use MPEG and QuickTime movies. Netscape is highly recommended.

# INFRARED SPACE OBSERVATORY

http://isowww.estec.esa.nl

The Infrared Space Observatory, a project of the European Space Agency, operates at wavelengths from 2.5 to 240 microns. The Infrared Space Observatory helps astronomers study and explore the objects of our universe with an instrument complement that includes an imaging photo-polarimeter, a short wavelength spectrometer, and a long wavelength spectrometer. To find out more about the project, stop off at this useful and informative Web site.

# LIMITING MAGNITUDE

http://www.funet.fi/pub/astro/html/eng/obs/rjm.html

Limiting magnitude is used to evaluate the quality of observing conditions in meteor and deep sky observations. It can be used also to approximate light pollution.

The simplest way to evaluate limiting magnitude is to find suitable stars with known magnitudes and check which of them are visible. A more complex method is to calculate visible stars inside known star squares and triangles including the corner stars. This method was originally used by meteor observers.

The Limiting Magnitude Web page provides a valuable reference to the twenty sky areas used in limiting magnitude estimation, their corner stars, and their constellations.

# MACHO PROJECT

http://wwwmacho.anu.edu.au

The MACHO Project is searching for dark matter in the form of Massive Compact Halo Objects (MACHOs) using gravitational microlensing. The project searches for these events by monitoring over 20 million stars every night in the LMC and Galactic bulge, using the dedicated 50-inch telescope at Mt. Stromlo. Find complete details on the project, as well as results to date, at this Web site.

# MULTIWAVELENGTH ATLAS OF GALAXIES

http://hea-www.harvard.edu/~mackie/atlas/atlas_edu.html

The Multiwavelength Atlas of Galaxies includes optical, x-ray, far-infrared, and radio images are shown for a variety of nearby galaxies. Text describing the physical mechanisms of the different types of radiation, and their astronomical sources, is supplied.

# NASA

http://hypatia.gsfc.nasa.gov/NASA_homepage.html

Care to take a cyberwalk through the Kennedy, Goddard, Ames, Dryden, and Langley space centers? Or would you like to let your fingers do the walking through some great NASA history files, the NASA Strategic Plan, and NASA's on-line educational resources? More to the point, you may just want to browse NASA's technical report server or the complete searchable index to NASA resources by subject.

# THE NINE PLANETS: A MULTIMEDIA TOUR

**http://seds.lpl.arizona.edu/billa/tnp**

This astonishing resource comprises a sixty-page hypermedia tour of the solar system that includes a marvelous collection of images, film-clips, and documentation.

# PARKES MULTIBEAM SURVEY

**http://wwwatnf.atnf.csiro.au/Research/**

The Parkes 64-m telescope is commencing an HI Southern Sky and Zone of Avoidance survey in 1996. The survey covers redshifts up to 0.04 and is sensitive to objects with HI mass between $10^6$ and $10^{10}$ solar masses, depending on distance. This is the first extensive "blind" survey of the 21-cm extragalactic sky.

# PROJECT GALILEO: BRINGING JUPITER TO EARTH

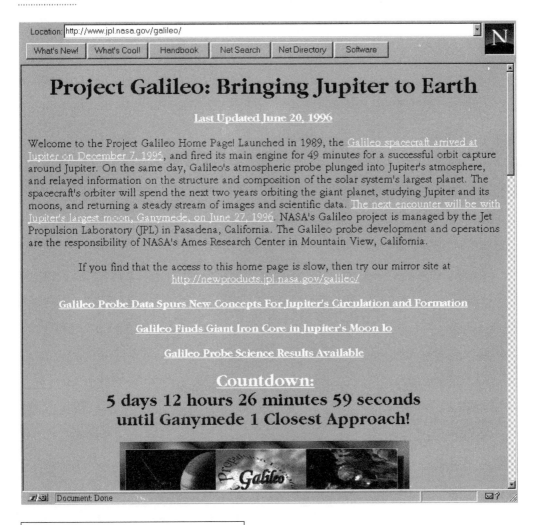

**http://www.jpl.nasa.gov/galileo**

The Galileo spacecraft arrived on Jupiter on December 7, 1995, and fired its main engine for 49 minutes for a successful orbit capture around the planet. On the same day, Galileo's atmospheric probe plunged into Jupiter's atmosphere, and relayed information on the structure and composition of the solar system's largest planet. Thus began what will be two years of Galileo's orbiting the giant planet. The output of images and scientific data regarding Jupiter and its moons has already been extensive.

Why is Galileo going to Jupiter when the two Voyagers have already been there? The Voyagers were like a quick car trip past the Grand Canyon—drive by, snap a few pictures, check off that you've been there and move on. Galileo is like stopping to explore the canyon, walking the trails (satellite tour), riding the rapids (Prove mission), and generally taking in the full majesty of the place (fields and particles).

Come to this extensive Web site for complete background information on Galileo, along with a complete gallery of images returned by the spacecraft, and much more. Here you will also find links to other home pages related to Galileo (including the home page for the Galileo Probe team at the Ames Research Center).

# REGIONAL ASTRONOMY CLUBS

Ames (Iowa) Area Amateur Astronomers
**http://www.cnde.iastate.edu/aaaa.html**

Astronomical Society of the Atlantic (Southeastern US)
**http://www.america.net/~erg/asa.html**

Eugene (Oregon) Astronomical Society
**http://www.efn.org/~bsackett/**

Finnish Amateur Astronomy Home Page
**http://www.funet.fi/pub/astro/html/astro-uk.html**

Orwell Astronomical Society (Ipswich, United Kingdom)
**http://www.ast.cam.ac.uk/~ipswich/**

Peoria (Illinois) Astronomical Society

**http://bradley.bradley.edu/~dware/index.html**

Prairie Astronomy Club (Lincoln, Nebraska)

**http://www.infoanalytic.com/pac/index.html**

Royal Astronomical Society of Canada

**http://apwww.stmarys.ca/rasc/rasc.html**

Student Astronomers of Harvard-Radcliffe: STAHR

**http://hcs.harvard.edu/~stahr**

United Kingdom Amateur Astronomy Society

**http://www.emoticon.com/astro**

# SERENDIP: SEARCH FOR EXTRATERRESTRIAL RADIO EMISSIONS FROM NEARBY DEVELOPED INTELLIGENT POPULATIONS

**http://albert.ssl.berkeley.edu/serendip**

SERENDIP, the UC Berkeley SETI program, is an ongoing scientific research effort aimed at detecting radio signals from extraterrestrial civilizations. The project is the world's only "piggyback" SETI system, operating alongside simultaneously conducted conventional radio astronomy observations. SERENDIP is currently piggybacking on the 1,000-foot dish at Arecibo Observatory in Puerto Rico, the largest radio telescope in the world.

# SETI INSTITUTE HOME PAGE

http://www.seti-inst.edu/

The SETI (Search for Extraterrestrial Intelligence) Institute serves as an institutional home for scientific and educational projects relevant to the nature, distribution, and prevalence of intelligent life in the universe. The largest research effort is Project Phoenix, a privately funded continuation of the Target Search portion of NASA's High Resolution Microwave Survey. Two other projects of interest are the Life in the Universe (LITU) Curriculum Project and the Flight Opportunities for Science Teacher EnRichment (FOSTER) Project. LITU develops supplementary science curriculum material for grades 3 through 9. FOSTER allows science teachers to experience the excitement of research on NASA's Kuiper Airborne Observatory (KAO).

# SKY ON-LINE

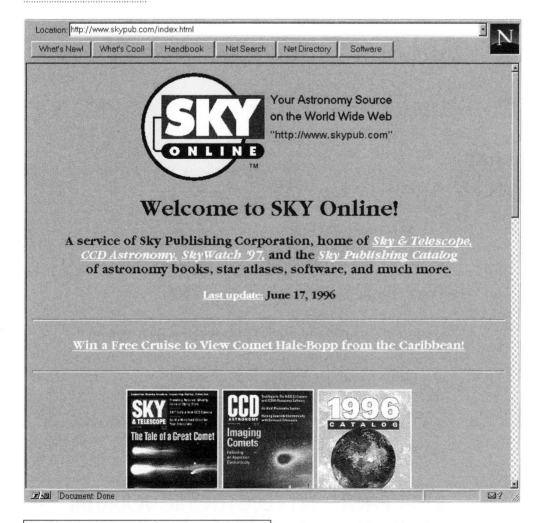

Location: http://www.skypub.com/index.html

| What's New! | What's Cool! | Handbook | Net Search | Net Directory | Software |

Your Astronomy Source
on the World Wide Web
"http://www.skypub.com"

## Welcome to SKY Online!

A service of Sky Publishing Corporation, home of *Sky & Telescope, CCD Astronomy, SkyWatch '97,* and the *Sky Publishing Catalog* of astronomy books, star atlases, software, and much more.

Last update: June 17, 1996

### Win a Free Cruise to View Comet Hale-Bopp from the Caribbean!

Document: Done

**http://www.skypub.com/index.html**

This splendid Web site comes to you courtesy of Sky Publishing, which also brings you such fine publications as *Sky & Telescope*, *CCD Astronomy*, and the *Sky Publishing Catalog*. Come here for *Sky & Telescope*'s Weekly News Bulletin of celestial events, tips for backyard astronomers, great astronomical software for your computer, reviews of scopes and accessories, and an excellent on-line directory of astronomy clubs and planetariums.

## SKYVIEW

http://skyview.gsfc.nasa.gov/

SkyView is a facility available over the Net that allows users to conveniently retrieve data from public all-sky surveys. The user enters the position and size of the region desired and the surveys wanted, and then the data are extracted and formatted for the user.

## SLOAN DIGITAL SKY SURVEY SCIENCE ARCHIVE

http://tarkus.pha.jhu.edu/ScienceArchive/welcome.html

The Sloan Digital Sky Survey is a project to survey a 10,000 square degree area on the Northern sky over a 5-year period. A dedicated 2.5-m telescope is specially designed to take wide field (3.3°) images. Find more details on the project, as well as some results to date, at this Web site.

## STAR FACTS: AN ELECTRONIC JOURNAL ABOUT THE UNIVERSE

http://ccnet4.ccnet.com/odyssey/

Peter Harris's *Star Facts* Web site features many of the same links you will find in other astronomy sites elucidated in these pages: news of Comet Hale-Bopp, images from the Hubble Space Telescope, and so on. But what sets Harris's site apart from the others is the excellent way in which he has sought out those Web documents of interest to astronomers that might, without a link here, remain obscure and unfindable. Thus one link takes you to the e-text of Charles Petit's *San Francisco Chronicle* article about the Bay Area astronomical team that, in January 1996, discovered two planets thirty-five light years from earth that appear to be the right temperature to support liquid water and perhaps life. Other "astro-jewel" reprints, as Harris calls them, include discussions of the Galileo Probe that are not available elsewhere.

# WEB NEBULAE

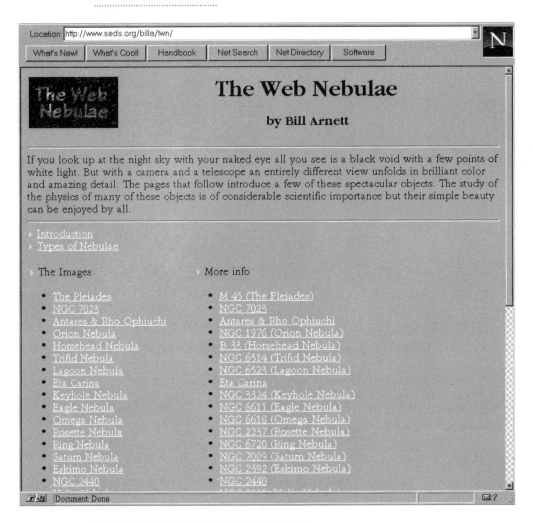

**http://www.seds.org/billa/twn**

The emphasis here is on aesthetics, not science. While there is a little information to help you appreciate the rich collection of images contained in the Web Nebulae, this is not a tutorial on the astronomy of nebulae. The creator of this site, Bill Arnett, writes, "If these images inspire you to study stellar evolution, that's great, but my intent is primarily to showcase a part of Nature's beauty that is accessible to us only via the telescope."

The main body of this Web document consists of two pages for each of about twenty-five objects. The first page consists almost entirely of a large image of the object. The second pages gives some basic information about the object and a list of references to other pictures of the object available on the Net. There are links at the bottom of each page to move to the next and previous objects, back and forth between the picture and description pages, and to the table of contents page. Useful appendices include a list of the sources from which these images were collected, an extensive list of images sorted by object, and an index and glossary.

The Web Nebulae is one of those Web sites that has been optimized for Netscape. It takes advantage of NetScape's newest features and therefore many of the images contained in the Web Nebulae. In particular, the Web Nebulae uses in-line JPEG images. This greatly reduces the time to load a page for those with slow Internet connections. The images are quite large after decompression, so you may find that Netscape works better if you give it a very large memory partition. The Web Nebulae will also "work" with MacMosaic and MacWeb, although the images won't be as clear as they are with Netscape.

The images include the Pleiades, NGC 7023, Antares and Rho Ophiuchi, the Orion Nebula, the Horsehead Nebula, the Trifid Nebula, the Lagoon Nebula, Eta Carina, the Keyhold Nebula, NGC 2440, Puppis A, the Vela Supernova Remnant, Supernova 1987a, and the Eagle, Omega, Rosette, Ring, Saturn, Eskimo, Helix, Dumbbell, Little Dumbbell, Cat's Eye, Crab, and Veil nebulae.

The creator of this site (Bill Arnett) is the same fellow we have to thank for *The Nine Planets* (above). And if you would in fact care to thank him, the e-mail address is billa@znet.com.

# WISCONSIN H-ALPHA MAPPER (WHAM)

http://wisp5.physics.wisc.edu/WHAM.html

The Wisconsin H-Alpha will produce a survey of H-Alpha emission from the interstellar medium (ISM) over the entire northern sky. The instrument combines a 24-inch telescope and a high-resolution, 6-inch Fabry-Perot spectrometer to achieve 8 to 12 km/s velocity resolution in 1° beams on the sky. This survey, the first of ionized hydrogen in our galaxy, will be used to explore the spatial and kinematic structure of the warm ionized component of the ISM. WHAM will move from the Pine Bluff Observatory to Kitt Peak in the fall of 1996 for the duration of the survey, which should take about two years.

# BIOLOGY

Alsbyte Biotech Products
American Association of Anatomists (AAA)
American Society of Plant Physiologists
Baylor Biological Databases
Bermuda Biological Station for Research (BBSR)
Biological Searches Page
Bio Online
BioMolecular Engineering Research Center (BMERC)
**Journal of Biological Chemistry**
**Caenorhabditis elegans** WWW Server
European Molecular Biology Laboratory
ExPASy
FlyBrain
Frontiers of Bioscience
Harvard Biological Laboratories
Hawke's Nest
Indiana University Bio Archive

Lawrence Berkeley National Laboratory Human Genome
    Center
Molecular Biology and BioInformations WWW Sampler
Molecular Biology Core Facilities/Dana-Farber Cancer
    Institute
Multiple Sequence Alignment Page
**Journal of Molecular Biology** Online
Laboratory of Patrick Nef
National Biological Service
Pedro's BioMolecular Research Tools
Polyfiltronics
Standards and Definitions for Molecular Biology Software
Society for the Study of Amphibians and Reptiles
Virology World Wide Web Server
The World Wide Web Journal of Biology
World Wide Web Sites of Biological Interest

# ALSBYTE BIOTECH PRODUCTS

http://www.alsbyte.com/bio/

Visit the Web home of Alsbyte Biotech Products for detailed information on:

❒ reagents for molecular and cellular biology—including enzymes and kits for molecular biology, monoclonal antibodies, and differential display kits for gene ID;

❒ software—including image analysis software for use with any scanner, Sequencher Software for contig alignment and DNA manipulation, vector NTI software for constriction and vector intelligent experiment design, and software for plasma drawing and sequence analysis.

# AMERICAN ASSOCIATION OF ANATOMISTS (AAA)

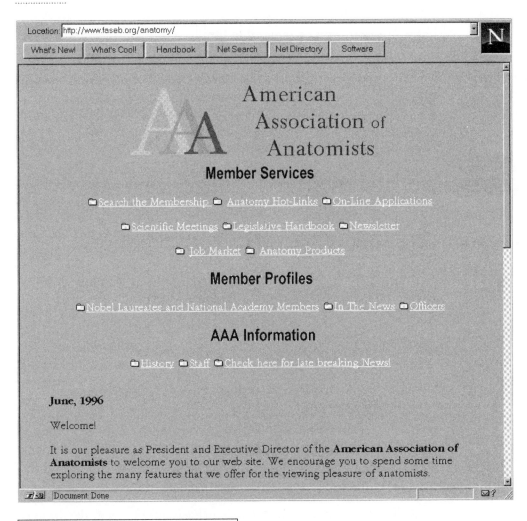

Location: http://www.faseb.org/anatomy/

| What's New! | What's Cool! | Handbook | Net Search | Net Directory | Software |

### American Association of Anatomists

## Member Services

Search the Membership  Anatomy Hot-Links  On-Line Applications

Scientific Meetings  Legislative Handbook  Newsletter

Job Market  Anatomy Products

## Member Profiles

Nobel Laureates and National Academy Members  In The News  Officers

## AAA Information

History  Staff  Check here for late breaking News!

**June, 1996**

Welcome!

It is our pleasure as President and Executive Director of the **American Association of Anatomists** to welcome you to our web site. We encourage you to spend some time exploring the many features that we offer for the viewing pleasure of anatomists.

Document Done

## http://www.faseb.org/anatomy

Here at the home page for the leading professional society for biological anatomists you will find everything from e-mail for AAA Nobel Laureates to classified job ads. You will also find information on joining the AAA.

# AMERICAN SOCIETY OF PLANT PHYSIOLOGISTS

http://www.aspp.org/member.htm

Visit the Web page of the premier professional society for plant scientists and access a wide range of information resources, including the three ASPP journals, *Plant Physiology*, *The Plant Cell*, and *The Plant Gene Register*.

# BAYLOR BIOLOGICAL DATABASES

http://mcbr.bcm.tmc.edu/MBCRdatabases.html

Here are three highly useful biological databases developed by the faculty of the Baylor College of Medicine.

The Mammary Transgene Interactive Database provides literature designed to target transgene proteins to the mammary gland. The current emphases are on biotechnology applications, tumor models, and development models.

The Small RNA Database presents a vast collection of information on RNAs not directly involved in protein synthesis. These are grouped under three categories: capped small RNAs, noncapped small RNAs, and viral small RNAs. Sequences and references are included, and you can do WAIS (Wide Area Information Server) searching by keyword.

Finally, the Tumor Gene Database is a fully relational database of genes associated with tumorigenesis and cellular transformation. This database includes oncogenes, proto-oncogenes, tumor suppressor genes/anti-oncogenes, regulators and substrates of the above, and other relevant genes and chromosomal regions.

# BERMUDA BIOLOGICAL STATION FOR RESEARCH (BBSR)

http://www.bbsr.edu/bbsr.html

BBSR's mission is threefold: To conduct research of the highest quality from the special perspective of a mid-ocean island, to educate future scientists, and to provide well-equipped facilities and technical staff support for visiting scientists, faculty, and students from around the world.

Come to this Web site to get information on BBSR's educational and scientific projects, as well as BBSR's initiative to create closer connections between the business and science communities.

# BIOLOGICAL SEARCHES PAGE

http://biomaster.uio.no/biological.html

The Biological Searches Page provides you with a wide range of Web and Internet search facilities related to biology, including:

❐ retrieval systems—including Genebank database searches

❐ sequence similarity searches—searches from Kabat's database and the Immuno Genetics database

❐ the BMC gene finder

❐ Pol3scan, which recognizes the eukaryotic internal control regions typical of tRNA genes and tDNA-derived genes

❐ protein sequence and motif database searches.

# BIO ON-LINE

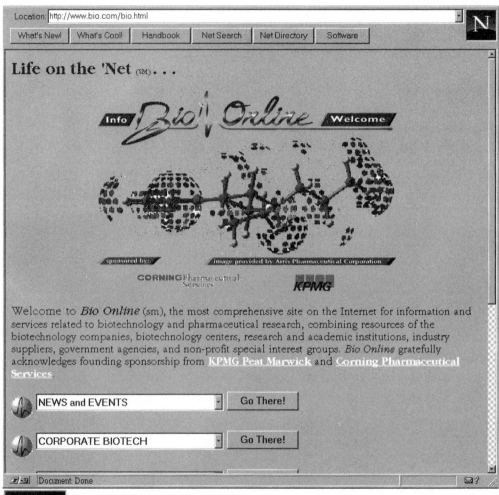

**META SITE**

**FREE STUFF**

**http://www.bio.com/bio.html**

Here is the most comprehensive site on the Internet for information and services related to biotechnology, combining resources of biotechnology companies, biotechnology centers, research and academic institutions, industry suppliers, government agencies, and nonprofit special interest groups.

# BIOMOLECULAR ENGINEERING RESEARCH CENTER (BMERC)

http://bmerc-www.bu.edu

The BioMolecular Engineering Research Center (BMERC) has two major research objectives:

❒ To develop statistical and other computational approaches that will detect syntactic and semantic patters in DNA, RNA, and protein sequences.

❒ To use statistical/computational approaches to identify structure, function, and regulation in these molecules. This identification has led to the formulation and testing of major hypotheses in the areas of molecular evolution, gene regulation, developmental genetics, and protein structure-function relationships.

The center's support program provides DNA, FNA, and protein sequence databases and analysis tools on-line to a large local area community and to the larger research community via its anonymous FTP and gopher servers. The center provides a distribution service of noncommercial software and support information for all developers, free of charge, to the scientific community as part of a larger dissemination program. In addition, the center has provided support for a number of interdisciplinary meetings and focuses on the computational challenges arising in molecular biology.

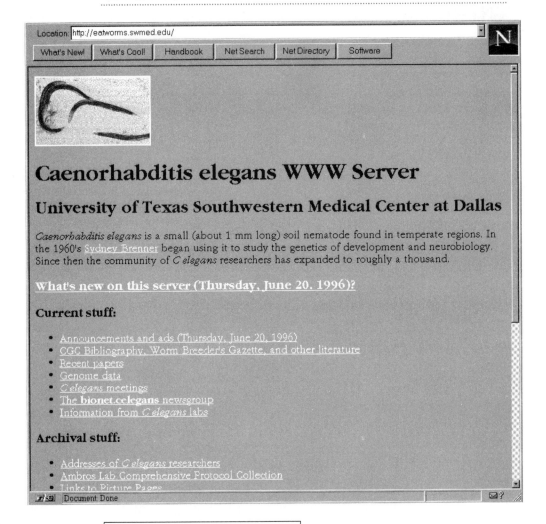

**Caenorhabditis elegans WWW Server**

**University of Texas Southwestern Medical Center at Dallas**

*Caenorhabditis elegans* is a small (about 1 mm long) soil nematode found in temperate regions. In the 1960's Sydney Brenner began using it to study the genetics of development and neurobiology. Since then the community of *C elegans* researchers has expanded to roughly a thousand.

**What's new on this server (Thursday, June 20, 1996)?**

**Current stuff:**

- Announcements and ads (Thursday, June 20, 1996)
- CGC Bibliography, Worm Breeder's Gazette, and other literature
- Recent papers
- Genome data
- *C elegans* meetings
- The **bionet.celegans** newsgroup
- Information from *C elegans* labs

**Archival stuff:**

- Addresses of *C elegans* researchers
- Ambros Lab Comprehensive Protocol Collection
- Links to Picture Pages

**http://eatworms.swmed.edu**

*Caenorhabditis elegans* is a small (about 1 mm long) soil nematode (worm) found in temperate regions. In the 1960s, Sydney Brenner began using it to study the genetics of development and neurobiology. Since then the community of *C. elegans* researchers has expanded greatly.

This Web site features a rich library of research papers in hypertext format, the latest research reports from various *C. elegans* labs, e-mail addresses for *C. elegans* researchers around the world, the complete archive of the Ambros Lab Comprehensive Protocol Collection, and more.

If you access this Web document and decide you like it, send your thanks and congrats to Leon Avery (Leon.Avery@email.swmed.edu).

# EUROPEAN MOLECULAR BIOLOGY LABORATORY

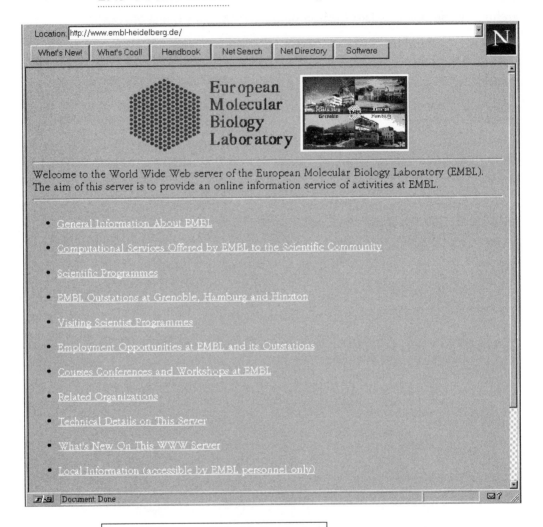

**http://www.embl-heidelberg.de/**

The European Molecular Biology Laboratory (EMBL) was established in 1974 and is supported by fourteen European countries and Israel. EMBL consists of a main laboratory in Heidelberg, and three outstations in Hamburg, Grenoble, and Hinxton (UK). As an institute with excellent facilities, EMBL prides itself on its intellectually open and challenging atmosphere, which encourages innovation and collaboration. The science performed here is first rate.

The current research activities of the main lab are carried out in four main research programs, each consisting of several independent groups focusing on a specific area of molecular and cell biology. Through its outstations in Hamburg and Grenoble, EMBL provides European scientists with access to highly intensive x-rays and neutron radiation for structural studies. The Hinxton outstation specializes in research and service in the field of bioinformatics.

EMBL is also a teaching and training center for the new generation of molecular biologists. Each year it organizes courses, symposia and workshops on current aspects of molecular biology.

Come to the EMBL Web site for more information on the various programs.

# EXPASY

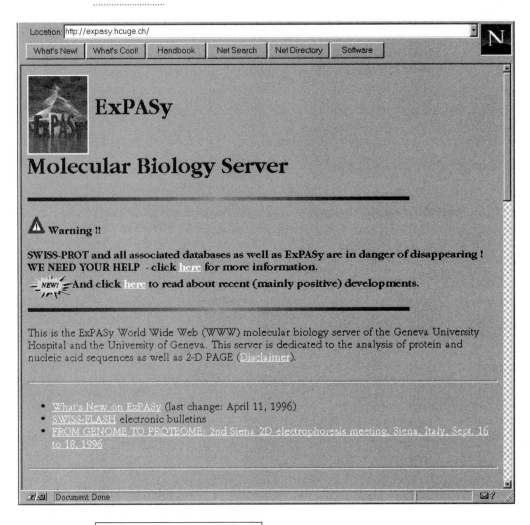

**http://expasy.hcuge.ch/**

The ExPASy molecular biology server of the Geneva University Hospital, University of Geneva, is dedicated to the analysis of protein and nucleic acid sequences. The Web site includes:

❏ an annotated protein sequence database

- ❐ an on-line atlas of protein sites and patterns

- ❐ a two-dimensional polyacrylamide gel electrophoresis (PAGE) database

- ❐ 3D images of proteins and other biological macromolecules

- ❐ a comprehensive enzyme nomenclature database

- ❐ and a sequence analysis bibliographic reference database.

The site also provides complete information and software for 2D PAGE analysis (including the Melanie II tutorial), an automated knowledge-based protein modeling server, and more.

# FLYBRAIN

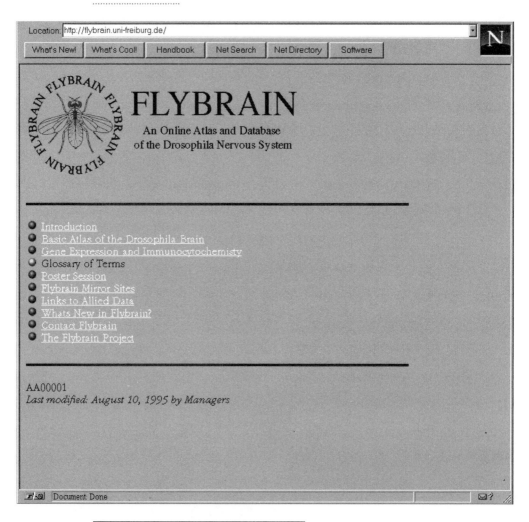

Location: http://flybrain.uni-freiburg.de/

| What's New! | What's Cool! | Handbook | Net Search | Net Directory | Software |

## FLYBRAIN

An Online Atlas and Database
of the Drosophila Nervous System

- Introduction
- Basic Atlas of the Drosophila Brain
- Gene Expression and Immunocytochemisty
- Glossary of Terms
- Poster Session
- Flybrain Mirror Sites
- Links to Allied Data
- Whats New in Flybrain?
- Contact Flybrain
- The Flybrain Project

AA00001
*Last modified: August 10, 1995 by Managers*

Document: Done

**http://flybrain.uni-frieberg.de/**

FlyBrain is an on-line atlas and database for the *Drosophila* nervous system. Come here for a variety of anatomical descriptions of the fly's central and peripheral nervous system with regard to both adults and individuals in other developmental stages. The site includes:

❏ Schematic representations providing diagrammatic views of the nervous system from a number of different perspectives—by clicking the cursor on a particular anatomical domain, you will embark on a tour that links textual descriptions to images obtained by a range of different visualization techniques.

❏ A silver stain atlas that depicts frontal, sagittal, and horizontal section series of the brain and thoracic/abdominal ganglia, displayed against a coordinate system. Specific structures (tracts, neuropils, cell body groups, receptor types) are labeled. Clicking the cursor on a particular label takes you to the corresponding entry in a glossary of anatomical terms, and/or allows movement to linked image files. Care to download an unannotated image for your own manipulation? Go right ahead.

❏ Golgi impregnations—a library of representative impregnations reveal neuronal architecture as serial sections, as 3D reconstructions, and as Quicktime/MPEG animations.

❏ Images of enhancer-trap lines that display unique aspects of the nervous system.

# FRONTIERS OF BIOSCIENCE

**http://www.allworld.com/bioscience**

This unique new on-line journal features articles of interest to professionals in any discipline in biology and medicine, including biochemistry, microbiology, parasitology, virology, immunology, biotechnology, and bioinformatics. The hypertext articles include rare images, videos, sounds, and assimilated data in the form of databases.

# HARVARD BIOLOGICAL LABORATORIES

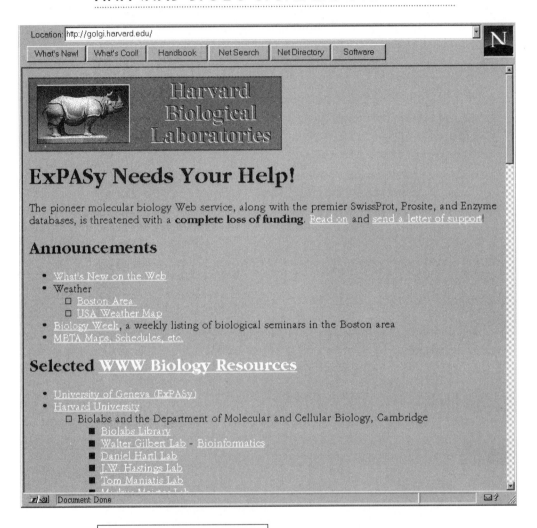

Location: http://golgi.harvard.edu/

What's New! | What's Cool! | Handbook | Net Search | Net Directory | Software

## Harvard Biological Laboratories

# ExPASy Needs Your Help!

The pioneer molecular biology Web service, along with the premier SwissProt, Prosite, and Enzyme databases, is threatened with a **complete loss of funding**. Read on and send a letter of support

## Announcements

- What's New on the Web
- Weather
  - Boston Area
  - USA Weather Map
- Biology Week, a weekly listing of biological seminars in the Boston area
- MBTA Maps, Schedules, etc.

## Selected WWW Biology Resources

- University of Geneva (ExPASy)
- Harvard University
  - Biolabs and the Department of Molecular and Cellular Biology, Cambridge
    - Biolabs Library
    - Walter Gilbert Lab - Bioinformatics
    - Daniel Hartl Lab
    - J.W. Hastings Lab
    - Tom Maniatis Lab

Document: Done

## http://golgi.harvard.edu

Here is access to the databases of Harvard's Biolabs Library, the Walter Gilbert lab (Bioinformatics), the Daniel Hartl lab, the J.W. Hastings lab, the Tom Maniatis lab, the Markus Meister lab, the Matt Meselson lab, the George M. Church lab, and the Roger Brent lab.

# HAWKE'S NEST

http://www.ucalgary.ca/~rhawkes/

Spatially repeated patterns in developing organisms—from segmentation in the leech through the patterned expression of homeodomain proteins in *Drosophila*, to somitic segregation in vertebrates—are a central problem in development. In the central nervous system, the mammalian cerebellum is an ideal tissue in which to explore pattern formation.

A variety of molecules—Zebrins—are expressed in the cerebellum of the adult mouse brain in an elegant array of stripes interposed by similar stripes of unlabeled cells. This pattern of stripes is, in turn, correlated closely with the patterns of axons bringing information into the cerebellum.

This raises four broad sets of issues. First, what does neuroanatomy tell us about the modular organization of the adult cerebellum? Second, how do cerebellar modules reflect the receptive fields and serve cerebellar function (motor control)? Third, how are the Zebrin bands generated during development, and how is expression directed to the Purkinje cell? Fourth, how do the different ingrowing climbing fiber and mossy fiber axons recognize their appropriate targets?

For approaches to all these questions, visit the Hawke's Nest Web site.

## INDIANA UNIVERSITY BIO ARCHIVE

http://iubio.bio.indiana.edu/

The Indiana University Bio Archive is a highly useful archive of biology data and software. The Archive includes items to browse, search, and fetch molecular data, software, biology news, and documents, as well as links to remote information sources in biology and related topics. One piece of particularly useful software available here is a great biosequencing editor for Macintosh, Microsoft Windows, and XWindows.

# JOURNAL OF BIOLOGICAL CHEMISTRY

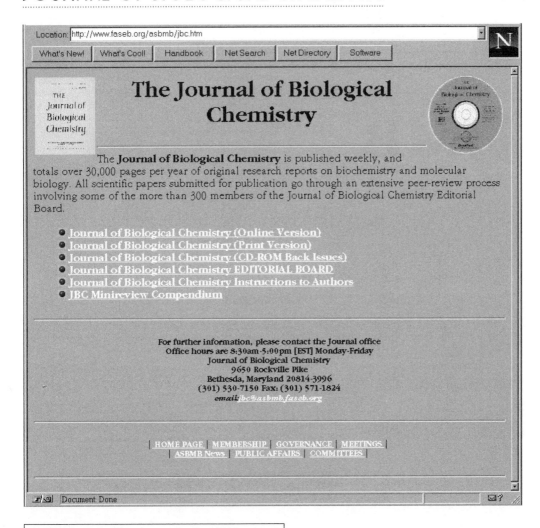

Location: http://www.faseb.org/asbmb/jbc.htm

What's New! | What's Cool! | Handbook | Net Search | Net Directory | Software

## The Journal of Biological Chemistry

The **Journal of Biological Chemistry** is published weekly, and totals over 30,000 pages per year of original research reports on biochemistry and molecular biology. All scientific papers submitted for publication go through an extensive peer-review process involving some of the more than 300 members of the Journal of Biological Chemistry Editorial Board.

- Journal of Biological Chemistry (Online Version)
- Journal of Biological Chemistry (Print Version)
- Journal of Biological Chemistry (CD-ROM Back Issues)
- Journal of Biological Chemistry EDITORIAL BOARD
- Journal of Biological Chemistry Instructions to Authors
- JBC Minireview Compendium

For further information, please contact the Journal office
Office hours are 8:30am-5:00pm [EST] Monday-Friday
Journal of Biological Chemistry
9650 Rockville Pike
Bethesda, Maryland 20814-3996
(301) 530-7150 Fax: (301) 571-1824
email jbc@asbmb.faseb.org

| HOME PAGE | MEMBERSHIP | GOVERNANCE | MEETINGS |
| ASBMB News | PUBLIC AFFAIRS | COMMITTEES |

Document: Done

## http://www.faseb.org/asbmb/jbc.htm

The *Journal of Biological Chemistry* is published by Academic Press and the American Society for Biochemistry and Molecular Biology, a nonprofit scientific and educational organization with over 9,000 members. This Web site features indices, TOCs, and abstracts for journal articles, together with information on how to subscribe.

# JOURNAL OF MOLECULAR BIOLOGY ON-LINE

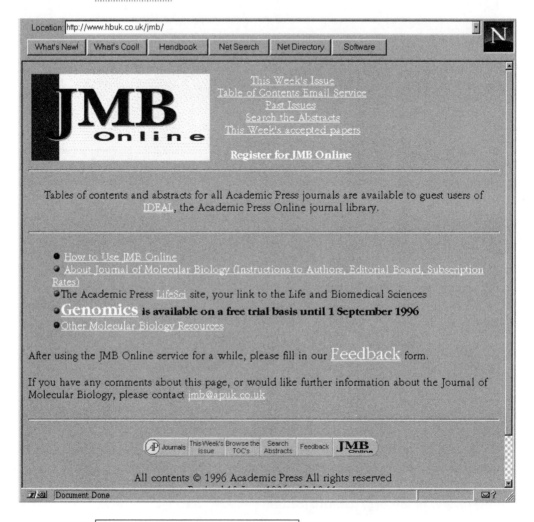

Location: http://www.hbuk.co.uk/jmb/

What's New! | What's Cool! | Handbook | Net Search | Net Directory | Software

This Week's Issue
Table of Contents Email Service
Past Issues
Search the Abstracts
This Week's accepted papers

**Register for JMB Online**

Tables of contents and abstracts for all Academic Press journals are available to guest users of IDEAL, the Academic Press Online journal library.

- How to Use JMB Online
- About Journal of Molecular Biology (Instructions to Authors, Editorial Board, Subscription Rates)
- The Academic Press LifeSci site, your link to the Life and Biomedical Sciences
- **Genomics is available on a free trial basis until 1 September 1996**
- Other Molecular Biology Resources

After using the JMB Online service for a while, please fill in our Feedback form.

If you have any comments about this page, or would like further information about the Journal of Molecular Biology, please contact jmb@apuk.co.uk

Journals | This Week's Issue | Browse the TOC's | Search Abstracts | Feedback | JMB Online

Document: Done

## http://www.hbuk.co.uk/jmb/

Established in 1959, the *Journal of Molecular Biology* is now available on-line on a subscription basis. Browse the contents of past issues, and search for and download abstracts and full texts of papers.

# LABORATORY OF PATRICK NEF

http://www.unige.ch/sciences/biochemie/Nef/nef.htm

Check into the lab of Patrick Nef and company at the Department of
Biochemistry, University of Geneva, where pioneering work is being
done on olfaction and neuronal calcium sensors.

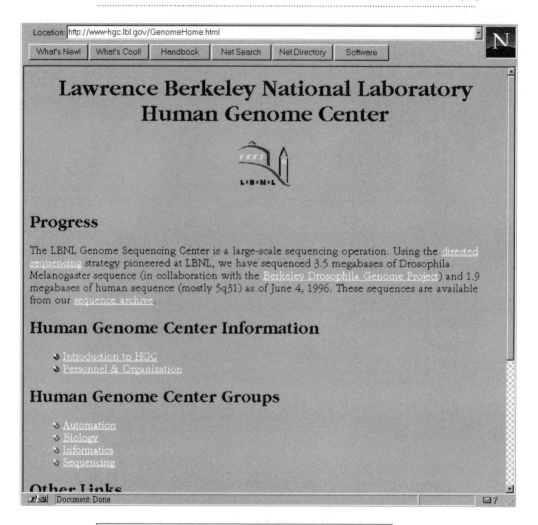

## Lawrence Berkeley National Laboratory Human Genome Center

**L·B·N·L**

### Progress

The LBNL Genome Sequencing Center is a large-scale sequencing operation. Using the directed sequencing strategy pioneered at LBNL, we have sequenced 3.5 megabases of Drosophila Melanogaster sequence (in collaboration with the Berkeley Drosophila Genome Project) and 1.9 megabases of human sequence (mostly 5q31) as of June 4, 1996. These sequences are available from our sequence archive.

### Human Genome Center Information

- Introduction to HGC
- Personnel & Organization

### Human Genome Center Groups

- Automation
- Biology
- Informatics
- Sequencing

### Other Links

**http://www-hgc.lbl.gov/GenomeHome.html**

The Lawrence Berkeley National Laboratory's Genome Sequencing Center has prepared 2.3 megabases of *Drosophila melanogaster* sequence and 700 kilobases of human 5q31. These sequences are now available from the FTP archive associated with this Web site. Here you will also find a complete guide to the personnel and organizations of the Human Genome Center, including the automation, biology, informatics, and sequencing groups.

# MOLECULAR BIOLOGY AND BIOINFORMATIONS WWW SAMPLER

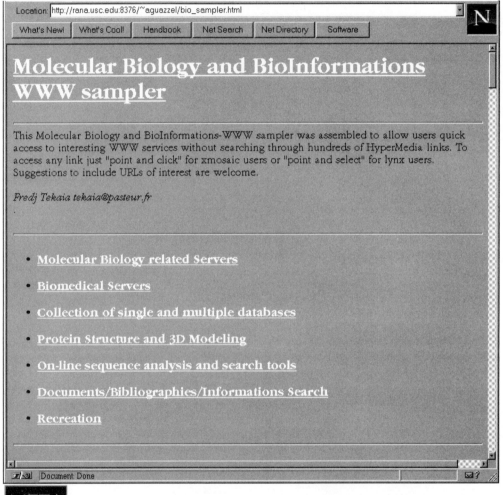

Location: http://rana.usc.edu:8376/~aguazzel/bio_sampler.html

What's New! | What's Cool! | Handbook | Net Search | Net Directory | Software

## Molecular Biology and BioInformations WWW sampler

This Molecular Biology and BioInformations-WWW sampler was assembled to allow users quick access to interesting WWW services without searching through hundreds of HyperMedia links. To access any link just "point and click" for xmosaic users or "point and select" for lynx users. Suggestions to include URLs of interest are welcome.

*Fredj Tekaia tekaia@pasteur.fr*

- **Molecular Biology related Servers**
- **Biomedical Servers**
- **Collection of single and multiple databases**
- **Protein Structure and 3D Modeling**
- **On-line sequence analysis and search tools**
- **Documents/Bibliographies/Informations Search**
- **Recreation**

Document: Done

http://rana.usc.edu:8376/~aguazzel/bio_sampler.html

This WWW sampler was assembled to allow users quick access to interesting WWW services without searching through hundreds of links. Here you have links to forty key molecular biology servers, eight major biomedical servers, as well as links to a wide range of single and multiple database servers, sites dealing with protein structure and 3D modeling, on-line sequence analysis and search tools, and hypertext bibliographies.

# MOLECULAR BIOLOGY CORE FACILITIES/DANA-FARBER CANCER INSTITUTE

http://mbcf. dfci.harvard.edu

Visit Dana-Farber's Web site to access myriad analysis services including:

❑ automated DNA sequencing—including single and double strand, and dye labeled dideoxyterminators

❑ DNA synthesis—40 nanomole, 0.2 micromole, 1 micromole, S-oligos, biotin, aminolink, dye labeled, and RNA oligos

❑ Peptide synthesis—amino acid analysis, analytical high-performance liquid chromatography (HPLC), and preparative HPLC

❑ Protein sequencing—including protein digest and mapping.

# MULTIPLE SEQUENCE ALIGNMENT PAGE

http://biomaster.uio.no/MSA.htm

Come here for benchmark references and software tools for the analysis of multiply aligned sequence, the calculation of multiple alignments, the calculation of phylogenetic trees, protein sequence alignment and database scanning, as well as link access to the PredictProtein server utilizing MaxHom alignments.

# NATIONAL BIOLOGICAL SERVICE

http://www.its.nbs.gov/nbs

The National Biological Service is working cooperatively with federal and state agencies to share information necessary to develop a comprehensive picture of the nation's biological resources. In the process, they have developed a highly useful database of such data, available at this site.

# PEDRO'S BIOMOLECULAR RESEARCH TOOLS

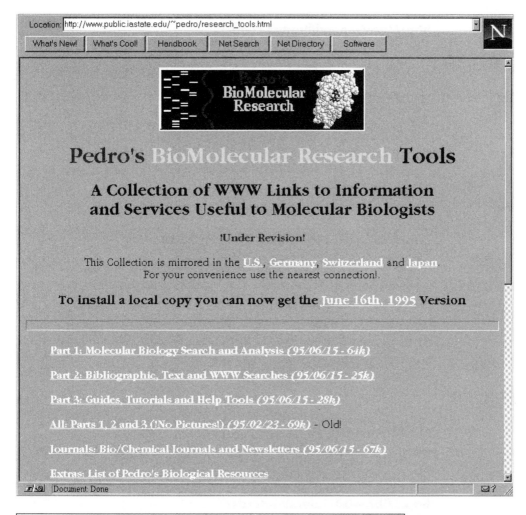

**Location:** http://www.public.iastate.edu/~pedro/research_tools.html

| What's New! | What's Cool! | Handbook | Net Search | Net Directory | Software |

**Pedro's** BioMolecular Research **Tools**

## A Collection of WWW Links to Information and Services Useful to Molecular Biologists

!Under Revision!

This Collection is mirrored in the U.S., Germany, Switzerland and Japan. For your convenience use the nearest connection!.

**To install a local copy you can now get the June 16th, 1995 Version**

Part 1: Molecular Biology Search and Analysis (95/06/15 - 64k)

Part 2: Bibliographic, Text and WWW Searches (95/06/15 - 25k)

Part 3: Guides, Tutorials and Help Tools (95/06/15 - 28k)

All: Parts 1, 2 and 3 (!No Pictures!) (95/02/23 - 69k) - Old!

Journals: Bio/Chemical Journals and Newsletters (95/06/15 - 67k)

Extras: List of Pedro's Biological Resources

Document: Done

**http://www.public.iastate.edu/~pedro/research_tools.html**

Here is a splendid collection of software tools for molecular biology search and analysis as well as bibliographic, text, and Web searches. As a bonus, access the comprehensive file of tutorials, images, journals, and newsletters.

# POLYFILTRONICS

http://www.polyfiltronics.com/

The home page of the company that makes Filter Bottom Microplates (FBMs), which have become perhaps the most essential of tools in drug screening and diagnostic arrays.

# SOCIETY FOR THE STUDY OF AMPHIBIANS AND REPTILES

http://falcon.cc.ukans.edu/~gpisani/SSAR.html

SSAR, a nonprofit organization established to advance research, conservation, and education concerning amphibians and reptiles, was founded in 1958. It is the largest international herpetological society, and is recognized worldwide for having an exceptional and diverse research program. Come to this home page for more information on SSAR.

# STANDARDS AND DEFINITIONS FOR MOLECULAR BIOLOGY SOFTWARE

http://ibc.wustl.edu/standards/

Come here for quick reference to CAS chemical exchange ASN.1 definitions (CXF-1.0), NCBI Bio-seq ASN.1 definitions, SCF DNA sequence file definitions, IUCr crystallographic information, the Draft 2 proposal for a Chemical MIME type, the PIR (codata) sequence format, and the Swiss-prot sequence format.

# VIROLOGY WORLD WIDE WEB SERVER

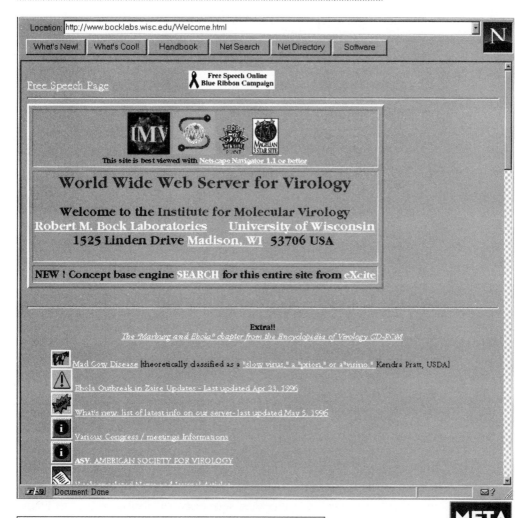

**http://www.bocklabs.wisc.edu/Welcome.html**

Maintained by the Institute for Molecular Virology at the Robert M. Bock Laboratories, University of Wisconsin, this wonderful Web site includes links to:

❐ the American Society for Virology

- ❐ digital reprints of virology-related news and journal articles

- ❐ computer visualizations of viruses

- ❐ topographical maps of viruses

- ❐ virus sequences, alignments and phylogenetic trees

- ❐ ICTV Classification of Viruses (both the fifth and sixth reports)

- ❐ course notes and tutorials on selected topics in general virology

- ❐ digitized images of viruses seen by electron microscopes

- ❐ hyperactive courses in virology (hypersyllabi!)

- ❐ a phone book of virologists on the Internet.

# WORLD WIDE WEB JOURNAL OF BIOLOGY

http://epress.com/w3jbio/

The *World Wide Web Journal of Biology* is an international monthly publication of Epress, Inc. Recent articles include discussion of inhibition of adventitious roots in mung bean cuttings, molecular dynamics simulations of glucocorticoid receptor protein in complex with a glucocorticoid response element, a stereochemical rationale for the genetic code derived from complementary fit of amino acids into cavities formed in codon/anticodon sequence of double stranded DNA, and more.

# WORLD WIDE WEB SITES OF BIOLOGICAL INTEREST

http://www.abc.hu/bios.tes.html

Here are links to hundreds of categorized tutorials, conference home pages, on-line documentation, biological databases, and servers of biological research institutes around the world.

# BIOSPHERE/ CONSERVATION/ ECOLOGY

Australian Environment
Biodiversity and Biological Collections Web Server
Biodiversity and Ecosystems Network (BENE)
Boyce Thompson Institute for Plant Research Environmental
     Biology Program
**Canadian Journal of Forest Research**
Ecological Resources Home Page of the Kennedy Space
     Center
Environmental Modeling and Visualization with GRASS
Environmental Sites on the Internet
FireNet
GAIA Forest Archives
Gap Analysis Software
Great Lakes Program of the State University of New York,
     Buffalo

Hudson River Sloop Clearwater
Imaging System Laboratory, University of Illinois
International Center for Tropical Ecology
The Lamont-Doherty Earth Observatory
Landscape Ecology and Biogeography
The Long-Term Ecological Research (LTER) Program
National Estuary Program of the EPA
Oak Ridge National Laboratory: Environmental Sciences
     Division
Silviculture Laboratory
Sustainable Forests Directory
Sylvan Display Program
Tropical Forest Research Group
WWW-Server for Ecological Monitoring
WWW Virtual Library of Forestry

# AUSTRALIAN ENVIRONMENT

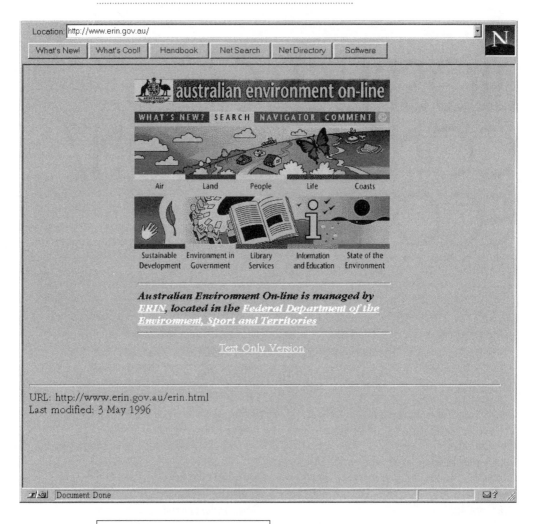

## http://www.erin.gov.au/

Here are all the references resources you need on the greenhouse effect and climate change in Australia, air pollution, terrestrial and inland aquatic landscapes, Australian biodiversity, and more.

# BIODIVERSITY AND BIOLOGICAL COLLECTIONS WEB SERVER

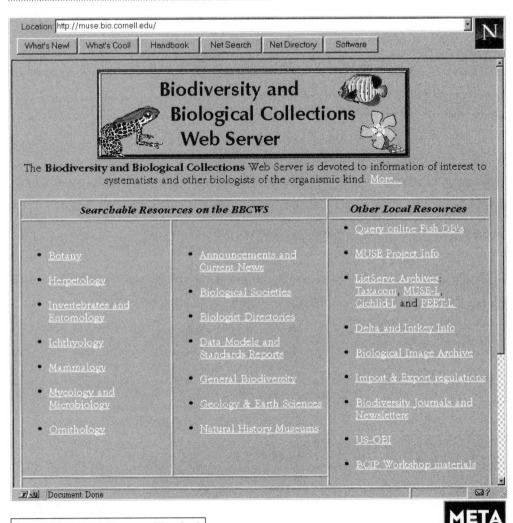

Location: http://muse.bio.cornell.edu/

What's New! | What's Cool! | Handbook | Net Search | Net Directory | Software

## Biodiversity and Biological Collections Web Server

The **Biodiversity and Biological Collections** Web Server is devoted to information of interest to systematists and other biologists of the organismic kind. More...

| Searchable Resources on the BBCWS | | Other Local Resources |
| --- | --- | --- |
| | | • Query online Fish DB's |
| • Botany | • Announcements and Current News | • MUSE Project Info |
| • Herpetology | | • ListServe Archives: Taxacom, MUSE-L, Cichlid-L and FEET-L |
| • Invertebrates and Entomology | • Biological Societies | |
| | • Biologist Directories | • Delta and Intkey Info |
| • Ichthyology | • Data Models and Standards Reports | • Biological Image Archive |
| • Mammalogy | | |
| • Mycology and Microbiology | • General Biodiversity | • Import & Export regulations |
| | • Geology & Earth Sciences | • Biodiversity Journals and Newsletters |
| • Ornithology | • Natural History Museums | |
| | | • US-OBI |
| | | • BCIP Workshop materials |

Document: Done

## http://muse.bio.cornell.edu/

The Biodiversity and Biological Collections Web Server is devoted to information of interest to systematists and other biologists of the organismic kind. The site provides hundreds of links to sources in botany, herpetology, invertebrates and entomology, ichthyology, mammalogy, mycology and microbiology, ornithology, and more. You also get resources in geology and the other earth sciences, a biological image archive, and connections to biodiversity journals and newsletters on-line.

# BIODIVERSITY AND ECOSYSTEMS NETWORK (BENE)

http://straylight.tamu.edu/bene/bene.html

The Biodiversity and Ecosystems Network (BENE) is designed to foster enhanced communications and collaborations among those interested in biodiversity conservation and ecosystem protection, restoration, and management. Here you will find hundreds of vital links to worldwide resources.

# BOYCE THOMPSON INSTITUTE FOR PLANT RESEARCH ENVIRONMENTAL BIOLOGY PROGRAM

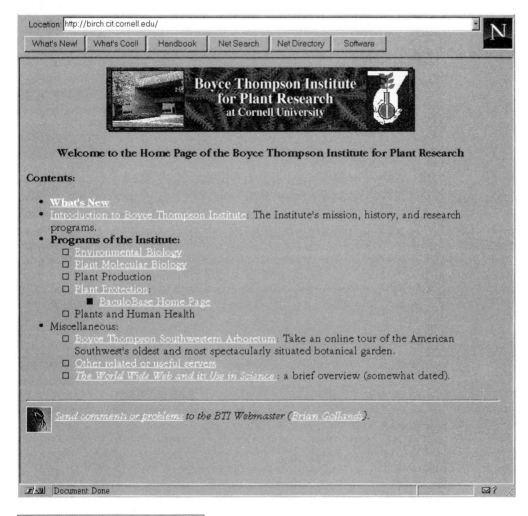

Location: http://birch.cit.cornell.edu/

| What's New! | What's Cool! | Handbook | Net Search | Net Directory | Software |

**Boyce Thompson Institute for Plant Research at Cornell University**

**Welcome to the Home Page of the Boyce Thompson Institute for Plant Research**

**Contents:**

- **What's New**
- Introduction to Boyce Thompson Institute: The Institute's mission, history, and research programs.
- **Programs of the Institute:**
  - Environmental Biology
  - Plant Molecular Biology
  - Plant Production
  - Plant Protection:
    - BaculoBase Home Page
  - Plants and Human Health
- Miscellaneous:
  - Boyce Thompson Southwestern Arboretum: Take an online tour of the American Southwest's oldest and most spectacularly situated botanical garden.
  - Other related or useful servers
  - The World Wide Web and its Use in Science : a brief overview (somewhat dated).

Send comments or problems to the BTI Webmaster (Brian Golland).

Document: Done

## http://birch.cit.cornell.edu/

Boyce Thompson's goal is to make this site a source of information for anyone interested in plants and the environment—from scientists and policy-makers to students. Come here for detailed information on the Environmental Biology Program's mission, the range of its research, personnel, facilities, and more. Come here also for a virtual tour of the Boyce Thompson Southwestern Arboretum, "the American Southwest's oldest and most spectacularly situated botanical garden."

# CANADIAN JOURNAL OF FOREST RESEARCH

http://www.cisti.nrc.ca/cisti/journals/

The *Canadian Journal of Forest Research* is published in English and French and has been ranked as one of the top journals in its field for the past decade by the Institute for Scientific Information (ISI). Over 65% of the articles published are from international (i.e., non-Canadian) sources. The journal offers research articles in silviculture, ecophysiology, forest ecology, biotechnology, forest genetics and tree improvement, tree physiology, forest entomology and pathology, land management and classification, wood processing, pollution effects, forest economics, and other forest related topics. Come to this useful Web page for journal TOCs, article submission guidelines, subscription information, and more.

# ECOLOGICAL RESOURCES HOME PAGE OF THE KENNEDY SPACE CENTER

http://atlas.ksc.nasa.gov/env.html

The Biomedical Operations and Research Office at the NASA John F. Kennedy Space Center (KSC) has been supporting environmental monitoring and research since the mid-1970s. Protection, preservation, and enhancement of the unique natural environment at KSC is a major objective of the center's strategic plan. Come to this dedicated Web page for information on KSC's research on vegetation, soil types, threatened and endangered species, jurisdictional wetlands, ground water recharge zones, surface water quality, climate, air quality, fire ecology, and more.

# ENVIRONMENTAL MODELING AND VISUALIZATION WITH GRASS

> http://www.cecer.army.mil/grass/viz/VIZ.html

Access a splendid collection of links that deliver visualizations created using GRASS (geographic resource analysis support system). These images are highly useful in the analysis of environmental models and spatial analysis algorithms and for data quality verification.

# ENVIRONMENTAL ORGANIZATIONS ON-LINE

Biodiversity Action Network
**http://www.access.digex.net/~bionet**

Byrd Polar Research Center
**http://www-bprc.mps.ohio-state.edu/**

Canadian Wildlife Federation
**http://www.toucan.net/cwf-fcf**

Chesapeake Bay Trust
**http://www2ari.net/home.cbt**

Citizen's Clearinghouse for Hazardous Waste
**http://www.essential.org/orgs/CCHW/CCHW.html**

Earth Pledge Foundation
**http://www.earthpledge.org/epfhome.html**

Earthlife Africa
**http://www.gem.co.za/ELA**

Earthwatch
**http://gaia.earthwatch.org**

EcoNews Africa
**http://www.web.apc.org/~econews/**

Ecotrust
**http://www.well.com/user/ecotrust/**

Friends of the Earth
**http://www.essential.org/FOE.html**

Global Rivers Environmental Education Network
**http://www.igc.apc.org/green/green.html**

Greenpeace
**http://www.greenpeace.org**

International Arid Lands Consortium
**http://ag.arizona.edu/OALS/IALC/Home.html**

League of Conservation Voters
**http://www.econet.apc.org/lcv/lcv_info.html**

National Audubon Society
**http://www.audubon.org**

National Environmental Information Resources Center
**http://www.gwu.edu/~greenu/**

National Parks and Conservation Association

**http://www.npca.com/pub/npca/**

National Wildlife Federation

**http://www.nwf.org/nwf/prog**

Natural Resources Defense Council

**http://www.igc.apc.org/nrdc**

Ozone Action

**http://www.essential.org/orgs/Ozone_Action/
Ozone_Action.html**

Ozone Depletion Over Antarctica

**http://icair.iac.org.nz/ozone/index.html**

Sierra Club

**http://www.sierraclub.org/**

The Wilderness Society

**http://town.hall.org/environment/wild_soc/wilderness.html**

World Conservation Monitoring Center

**http://www.wcmc.org.uk/**

World Wildlife Fund

**http://www.envirolink.org/orgs/wqed/wwf/wwf_home.html**

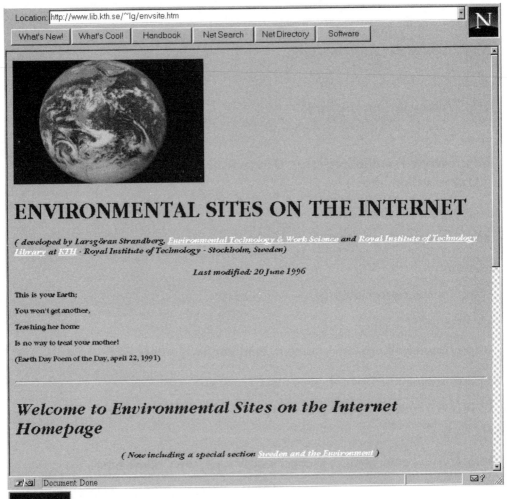

Location: http://www.lib.kth.se/~lg/envsite.htm

| What's New! | What's Cool! | Handbook | Net Search | Net Directory | Software |

# ENVIRONMENTAL SITES ON THE INTERNET

*( developed by Larsgöran Strandberg, Environmental Technology & Work Science and Royal Institute of Technology Library at KTH - Royal Institute of Technology - Stockholm, Sweden)*

*Last modified: 20 June 1996*

This is your Earth;

You won't get another,

Trashing her home

Is no way to treat your mother!

(Earth Day Poem of the Day, april 22, 1991)

## *Welcome to Environmental Sites on the Internet Homepage*

*( Now including a special section Sweden and the Environment )*

Document Done

**http://www.lib.kth.se/~lg/envsite.htm**

Covering everything from acid rain to whaling, from acoustic ecology to wetlands, this exhaustive resource gives you hundreds of links to resources on air and water pollution, nuclear radiation, ecological economics, global change, oil spills, pesticides, and much, much more.

# FIRENET

http://online.anu.edu.au/Forestry/fire/firenet.html

FireNet is an on-line information service for anyone interested in rural and landscape fires. The information contained here addresses all aspects of fire science and management including fire behavior, fire weather, fire prevention, mitigation and suppression, and plant and animal responses to fire and all aspects of fire effects.

# GAIA FOREST ARCHIVES

http://forests.lic.wisc.edu/forests/gaia.html

Here are links to more than 2,000 home pages of forest/biodiversity information. The database is broken down by country and provides significant information on worldwide forest protection campaigns and research over the past five years. You may sign up to have Biodiversity Campaign news e-mailed to you periodically on a free subscription basis.

# GAP ANALYSIS SOFTWARE

http://www.nr.usu.edu/gap/

Gap Analysis is a proactive approach to the management of biological diversity. The process involves the assessment of our current biological resources at the landscape level. Come here for a Gap Analysis "how-to" as well as for a free software download.

## GREAT LAKES PROGRAM OF THE STATE UNIVERSITY OF NEW YORK, BUFFALO

> http://wings.buffalo.edu/glp/

The mission of the Great Lakes Program is to develop, evaluate, and synthesize scientific and technical knowledge on the Great Lakes ecosystem in support of public education and policy formation.

## HUDSON RIVER SLOOP CLEARWATER

> http://clearwater.org

Since 1969, the Sloop *Clearwater* has sailed the Hudson River in an ongoing effort to foster cleaner waters via education, community action, and research. She sports the largest mainsail in the world, the most graceful of classic sloop designs, and the most noble of missions. To find out about the *Clearwater*'s programs, including how to arrange to sign on as a volunteer crew member for one week of estuary research and education on board, visit the sloop's home page. (In the interest of full disclosure, I admit that I have a bias here—I am a former member of the board of directors of the *Clearwater*.)

## IMAGING SYSTEM LABORATORY, UNIVERSITY OF ILLINOIS

> http://imlab9.landarch.uiuc.edu/

Access the home page of the Imaging Systems Laboratory of the Department of Landscape Architecture, University of Illinois, Champaign-Urbana. The lab was founded in 1986 to conduct environmental perception research using a variety of computer-based technologies. Most of the work at the lab has been in the visualization and evaluation of large-scale environmental changes—forest management, forest pests and fire, and harvesting practices—with emphasis on methodological issues. The lab was a pioneer in the use of computer image editing as a research tool, and continues to lead in integration of photo-realistic rendering, numerical modeling, and remote sensing for landscape analysis.

# INTERNATIONAL CENTER FOR TROPICAL ECOLOGY

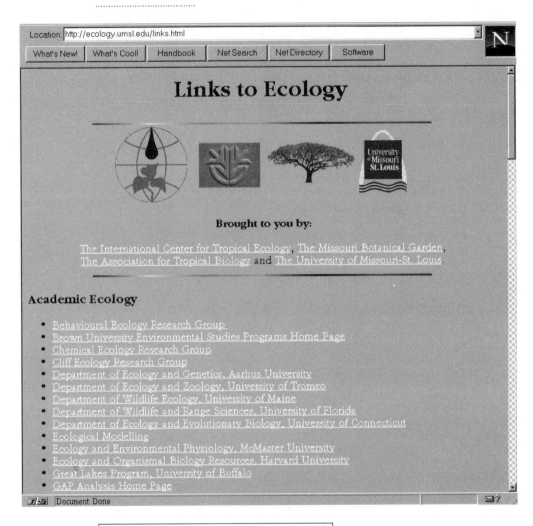

Location: http://ecology.umsl.edu/links.html

What's New! | What's Cool! | Handbook | Net Search | Net Directory | Software

## Links to Ecology

**Brought to you by:**

The International Center for Tropical Ecology, The Missouri Botanical Garden, The Association for Tropical Biology and The University of Missouri-St. Louis

### Academic Ecology

- Behavioural Ecology Research Group
- Brown University Environmental Studies Programs Home Page
- Chemical Ecology Research Group
- Cliff Ecology Research Group
- Department of Ecology and Genetics, Aarhus University
- Department of Ecology and Zoology, University of Tromso
- Department of Wildlife Ecology, University of Maine
- Department of Wildlife and Range Sciences, University of Florida
- Department of Ecology and Evolutionary Biology, University of Connecticut
- Ecological Modelling
- Ecology and Environmental Physiology, McMaster University
- Ecology and Organismal Biology Resources, Harvard University
- Great Lakes Program, University of Buffalo
- GAP Analysis Home Page

Document: Done

**http://ecology.umsl.edu/links.html**

Here are more than seventy links to, among other things, home pages for the Tropical Web in Brazil, resources in palynology and paleoclimatology, the Cooperative Research Center for Freshwater Ecology, and the University of Virginia's EcoWeb Project.

# LAMONT-DOHERTY EARTH OBSERVATORY

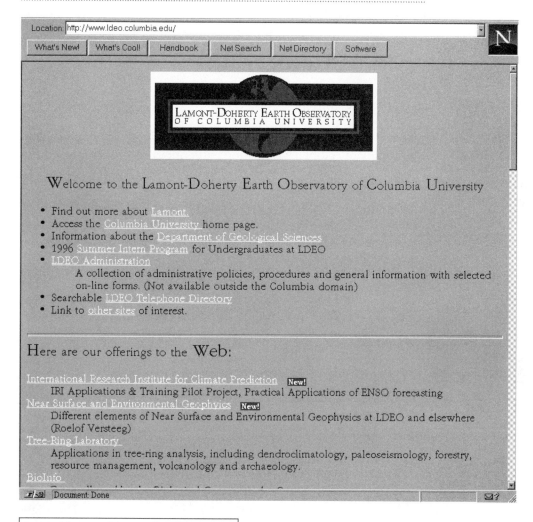

Location: http://www.ldeo.columbia.edu/

| What's New! | What's Cool! | Handbook | Net Search | Net Directory | Software |

LAMONT-DOHERTY EARTH OBSERVATORY
OF COLUMBIA UNIVERSITY

Welcome to the Lamont-Doherty Earth Observatory of Columbia University

- Find out more about Lamont.
- Access the Columbia University home page.
- Information about the Department of Geological Sciences
- 1996 Summer Intern Program for Undergraduates at LDEO
- LDEO Administration
    A collection of administrative policies, procedures and general information with selected on-line forms. (Not available outside the Columbia domain)
- Searchable LDEO Telephone Directory
- Link to other sites of interest.

Here are our offerings to the Web:

International Research Institute for Climate Prediction **New!**
    IRI Applications & Training Pilot Project, Practical Applications of ENSO forecasting
Near Surface and Environmental Geophysics **New!**
    Different elements of Near Surface and Environmental Geophysics at LDEO and elsewhere (Roelof Versteeg)
Tree-Ring Labratory
    Applications in tree-ring analysis, including dendroclimatology, paleoseismology, forestry, resource management, volcanology and archaeology.
BioInfo

Document: Done

## http://www.ldeo.columbia.edu/

The Lamont-Doherty Earth Observatory of Columbia University provides a wonderful Web resource comprising links to information on the range of lab programs as well as links to other, related institutions around the globe.

# LANDSCAPE ECOLOGY AND BIOGEOGRAPHY

http://life.anu.edu.au/landscape_ecology/landscape.html

This exhaustive and highly useful collection provides links to:

❏ landscape and environment resources—paleoenvironments, pollen, climate, global monitoring, greenness indices, plant taxonomy, and biodiversity

❏ geographic information and systems—the Australian Key Centre of Land Information Studies, the Geographic Information and Analysis Laboratory, Project GeoSim (Geography Education Software), and more.

# LONG-TERM ECOLOGICAL RESEARCH (LTER) PROGRAM

http://lternet.edu

With an initial set of six sites selected in 1980, the National Science Foundation established the Long-Term Ecological Research (LTER) Program to support research on long-term ecological phenomena in the United States. The present program consists of eighteen sites representing diverse ecosystems and research emphases. The LTER Network is a collaborative effort among over 775 scientists and students that extends the opportunities and capabilities of the individual sites to promote synthesis and comparative research across sites and ecosystems and among other related domestic and international research programs.

# NATIONAL ESTUARY PROGRAM OF THE EPA

**http://www.epa.gov/nep/nep.html**

Quite simply: the ultimate metasite on estuarine matters in the United States. There is no relevant link that you won't find here.

# OAK RIDGE NATIONAL LABORATORY: ENVIRONMENTAL SCIENCES DIVISION

**http://www.esd.ornl.gov/**

The Environmental Sciences Division (ESD) of the Oak Ridge National Laboratory is a division of that institution's Life Science and Environmental Technologies Directorate. The ESD is a multidisciplinary research and development organization with a staff of 210 plus more than 300 visiting professional collaborators and students. ESD's mission is to understand and evaluate how the development and use of energy affect the environment. To accomplish this mission, division staff conduct basic and applied research, assess environmental impacts of projects and policies, develop and demonstrate environmental technologies, and support educational activities. The unique combination of basic and applied research activities conducted in aquatic and terrestrial environments on the 550-acre Oak Ridge National Environmental Research Park makes the division one of the leading environmental research organizations in the United States.

# SILVICULTURE LABORATORY

## http://silvae.cfr.washington.edu

Brought to you by the good people at the College of Forest Resources, University of Washington, Seattle, the home page for the Silviculture Laboratory provides detailed resources relating to landscape management, as well as access to two highly useful pieces of software for IBM PCs.

The first of these software applications is the Stand Visualization System (SVS), which generates graphic images depicting stand conditions represented by a list of individual stand components, e.g., trees, shrubs, and down material. The images produced by SVS, while abstract, provide a readily understandable representation of stand conditions. Images produced using SVS help communicate silvicultural treatments and forest management alternatives to a variety of audiences.

The second piece of software, called UTOOLS Landscape Analysis Software, provides geographic analysis for watershed-level planning. The system provides a flexible framework for spatial analyses and can be used to address a variety of problems. The difference between UTOOLS and other spatial analysis software packages is that in UTOOLS all spatial data for a given project is integrated into a single Paradox database, where basic data operations can be quickly and easily performed. For instance, complex overlay operations that involve combinations of map layers and attributes can be done (and redone) with simple Paradox queries. Generating new layers from existing ones is also easy.

# SUSTAINABLE FORESTS DIRECTORY

> **http://www.together.net/~wow/Index**

The Sustainable Forests Directory is a project of the Forest Partnership, a nonprofit educational organization based in Burlington, VT. The mission of the Forest Partnership is to help bridge the gap between economic needs of the forest products industry and the ecological concerns of the environmental community. The directory provides dozens of links to vital forestry research sites nationwide, as well as reprints of important journal articles and information on relevant scientific meetings.

# SYLVAN DISPLAY PROGRAM

> **http://silva.snr.missouri.edu/display/sylvan.display.html**

The Sylvan Display Program is software for IBM PCs and compatibles designed to display a file that includes a list of trees in a stylized manner on a computer screen. It was originally developed to display the output of the Sylvan silvicultural management model—thus the name. Grab a free download of the software here.

# TROPICAL FOREST RESEARCH GROUP

> **http://ifs.plants.ox.ac.uk/OFI/tfrg/partic.htm**

The Tropical Forest Research Group (TFRG) is a voluntary association of institutions and organizations based in the south of England that have a demonstrated capacity in tropical forestry research, project management and consultancy. The consortium exists to foster collaboration and to mobilize the resources of the thirteen member organizations. At this Web site you will find not only information on TFRG's various programs, but also links to the home pages of its member organizations including Fountain Renewable Resources, the Natural Resources Institute, the Overseas Development Institute, the Oxford Forestry Institute, Silsoe College, the University of Reading, the University of Wales, and the World Conservation Monitoring Center.

## WWW-SERVER FOR ECOLOGICAL MONITORING

http://dino.wiz.uni-kassel.de/ecobas.html

This WWW-server provides easy access to available information about ecological modeling (simulation models, descriptions of these models, simulation software, people, and literature). Additionally, the server is meant to be a distribution tool for those modelers who wish to make their models easily available to others. For better utility, the server integrates an interface to ECOBAS Documentation of mathematical formulations for ecological processes.

## WWW VIRTUAL LIBRARY OF FORESTRY

http://www.metla.fi/info/vlib/Forestry.html

The WWW Virtual Library of Forestry provides more than 100 resources including working groups and networks, journals and newsletters, conference proceedings, mailing lists and Usenet newsgroups, bibliographies, research papers and other publications, legislation and international agreements, and even free downloadable forest modeling software.

# CHAOS AND FRACTALS

Applied Chaos Laboratory, Georgia Tech
The Beauty of Chaos
Bourke's Fractals Page
Chaos: The Course
The Chaos Game
Chaos at Maryland
The Chaos Metalink
Chaos Group at the S-DALINAC
Complexity International
Fractals and their Application to Geometry Models
Fractal Clouds
Fractal FAQ
Fractal Image Encoding
Fractals Machine
Fractal Microscope
Fractal Movie Archive
Fractals and Scale: A Tutorial
The Fractint Par Exchange
Fractint World Wide Web Pages
Fractal Video Art Gallery

Having Fun with Hydra
Iterated Function Systems Playground
Gallery of Interactive Geometry
Building Trees Using L-Systems
Mandelbrot Exhibition of the Virtual Museum of Computing
Mandelbrot Explorer
Mandelbrot and Julia Set Explorer
The Mandelbrot Set: A Java-Based Mandelbrot Explorer!!!!!!!
Mathart.com Fractal MUD
Mitsubishi Fractal Image Compression
Software for Nonlinear Dynamical Systems
**Nonlinear Science Today** and **The Journal of Nonlinear Science**
The Nonlinearity and Complexity Home Page
Clifford A. Pickover Home Page
The Spanky Fractal Database
Test Fractal Generator
VRML Fractals
Waterloo Montreal Verona Fractal Research Initiative
What Is a Fractal?

# APPLIED CHAOS LABORATORY, GEORGIA TECH

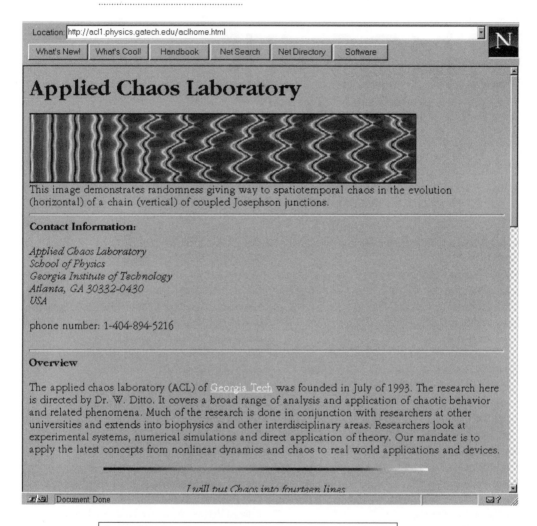

**http://acl1.physics.gatech.edu/aclhome.html**

The Applied Chaos Laboratory (ACL) at Georgia Tech was founded in July 1993. The research here is directed by Dr. W. Ditto and covers a broad range of analysis and application of chaotic behavior and related phenomena. Much of the research is done in conjunction with researchers at other universities and extends into biophysics and other interdisciplinary areas. Researchers look at experimental systems, numerical simulations, and direct application of theory. The lab's mandate is to apply the latest concepts from nonlinear dynamics and chaos to real world applications and devices.

# THE BEAUTY OF CHAOS

http://i30www.ira.uka.de/~ukrueger/fractals

Make a journey into the depths of the Mandelbrot set and discover the Beauty of Chaos using this database of over precalculated 500 images, many of them available for you to download.

# BOURKE'S FRACTALS PAGE

http://www.auckland.ac.nz/arch/pdbourke/fractals.html

Maintained by Paul Bourke of Auckland, New Zealand, this excellent Web page includes a wonderful HTML introduction to fractal science as well as free access to several extraordinary software programs that are available for download:

❏ FDC2D, a Macintosh application for calculating the fractal dimension of a PICT image

❏ FDC3D, a Mac application for calculating the fractal dimension of 3D geometry

❏ FracHill, a Mac application that generates fractal landscapes, renders them, and exports into lots of 3D formats including Radiance, RayShade, PV-Ray, and DXF

❏ LSystems—the full implementation from Lindenmayer Systems

❐ Fractal, which implements several different fractal/chaotic image generators.

## BUILDING TREES USING L-SYSTEMS

http://hill.lut.ac.uk/TestStuff/trees/

This site provides a number of trees produced using L-Systems. The program that created them was written on a PC by Phil Drinkwater at Loughborough University of Technology. In addition to the images, Drinkwater also provides you with a copy of the trees program itself.

## CHAOS: THE COURSE

http://www.lib.rmit.edu.au/fractals/exploring.html

Here is a wonderful, literate, clearly explained HTML primer on the rudiments of chaos. "Do you need to be told about chaos, or is your desk a permanent example?" it asks. "As everyone knows, beneath what those intolerably neat and tidy people consider to be chaos, there is a form of order. The chaotic housekeepers can always find the item of their desire—as long as no one tidies up! Many systems that scientists have considered totally random, unpredictable and without form have now been found to be otherwise." And so on.

## THE CHAOS GAME

http://math.bu.edu/DYSYS/chaos-game/node1.html

One of the most interesting fractals arises from what Michael Barnsley has dubbed "The Chaos Game." The Chaos Game is played as follows. First pick three points at the vertices of a triangle (any triangle works—right, equilateral, isosceles, whatever). Color one of the vertices red, the second blue, and the third green.

Next, take a die and color two of the faces red, two blue, and two green. Now start with any point in the triangle. This point is the seed for the game. (Actually, the seed can be anywhere in the plane, even miles away from the triangle.) Then roll the die. Depending on what color comes up, move the seed half the distance to the appropriately colored vertex. That is, if red comes up, move the seed half the distance to the red vertex. Now erase the original point and begin again, using the result of the previous roll as the seed for the next. That is, roll the die again and move the new point half the distance to the appropriately colored vertex, and then erase the starting point.

Keep on going. The resulting image is anything but a random smear. In fact the points will form what mathematicians call the Sierpinski triangle: one of the most basic types of geometric images known as fractals.

I've told you how the game will end, but it still might be something you'd find to be fun.

# CHAOS GROUP AT THE S-DALINAC

http://linac.ikp.physik.th-darmstadt.de/heiko/chaosmain.html

Check in with Germany's leading quantum chaos research group located at the Institut Für Kernphysik, Darmstadt. Here you will find biographies and bibliographies for each member of the group, along with the complete texts of research papers and other publications.

# CHAOS AT MARYLAND

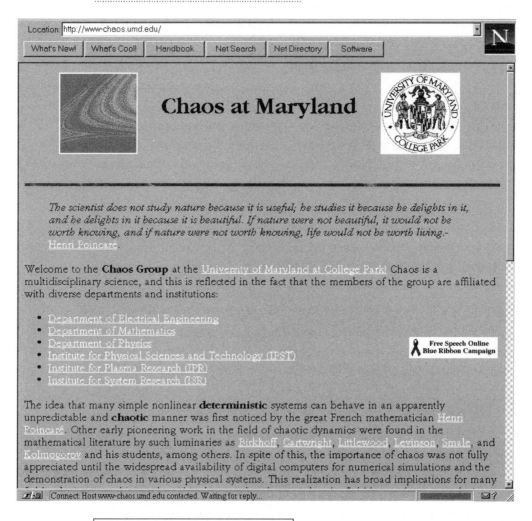

http://www-chaos.umd.edu/

Here is the home page of the Chaos Group at the University of Maryland, College Park. Since the mid-1970s the Chaos Group at Maryland has done extensive research in various areas of chaotic dynamics ranging from the theory of dimensions, fractal basin boundaries, chaotic scattering, and controlling chaos. Access the results of this research here.

# THE CHAOS METALINK

| http://www.industrialstreet.com/chaos/metalink.htm |

Comprising hundred of links, this site is designed as an ongoing resource for all those interested in chaos science and things fractal. The site is sponsored by Images of Chaos and Industrial Street Productions as a service to the chaos community. Here you have links to pages for software downloads, research abstracts, academic departments, and much more. This is a "must" item for the browser hotlist of anyone interested in chaos and fractals.

# COMPLEXITY INTERNATIONAL

| http://www.csu.edu/au |

*Complexity International* is a refereed electronic journal for scientific papers dealing with any area of complex systems research. The theme of the journal is the field of complex systems, the generation of complex behavior from the interaction of multiple parallel processes. Relevant topics include (but are not restricted to), artificial life, cellular automata, chaos theory, control theory, evolutionary programming, fractals, genetic algorithms, information systems, neural networks, nonlinear dynamics, and parallel computation.

# FRACTAL CLOUDS

http://climate.gsfc.nasa.gov/%7Ecahalan/FractalClouds/
FractalClouds.html

Come here for information on fractal cloud types, associated geometry, processes, and environment. You can also access appropriate models for various types of clouds as well as JPEG movies for zooming in on various clouds.

# FRACTAL FAQ

http://www.cis.ohio-state.edu/hypertext/faq/usenet/fractal-
faq/faq.html

The international computer network Usenet contains discussions on a variety of topics. The Usenet newsgroup sci.fractals and the listserve forum frac-1 are devoted to discussions on fractals. This FAQ (Frequently Asked Questions) list is an electronic serial compiled from questions and answers contributed by many participants in those discussions. This FAQ also lists various archives of programs, images, and papers that can be accessed through the global computer networks (WWW, Internet, and BITNET) by using e-mail, anonymous ftp, gophers, and World Wide Web browsers. This FAQ is not intended as a general introduction to fractals, or as a set of rigorous definitions, but rather as a useful summary of ideas, sources, and references. Unlike the version of this FAQ that is routinely posted to newsgroups, this Web hypertext version of the FAQ features links to related sources on the Web.

# FRACTAL IMAGE ENCODING

| http://inls.ucsd.edu/y/Fractals |

This site contains links to a variety of information and resources on fractal image encoding and related topics, including bibliographies, research papers rendered in HTML, and software.

# FRACTAL MICROSCOPE

| http://www.ncsa.uiuc.edu/Edu/Fractal/Fractal_Home.html |

Fractal Microscope combines supercomputing networks with the simple Mac or X-Window interface to create a powerful interactive tool for exploring fractal patterns. The program is meant to run with the NCSA imaging tools DataScope and Collage. Access Fractal Microscope here.

# FRACTAL MOVIE ARCHIVE

| http://www.cnam.fr/fractals/anim.html |

Here are more than 147 fractal animations available in Anim5, FLI, FLC, MPEG, and QuickTime formats. The animations include snow flakes in a red Mandelbrot set, galactic clusters, a Mandelbrot set morphing into a Julia set, and much more.

# FRACTAL VIDEO ART GALLERY

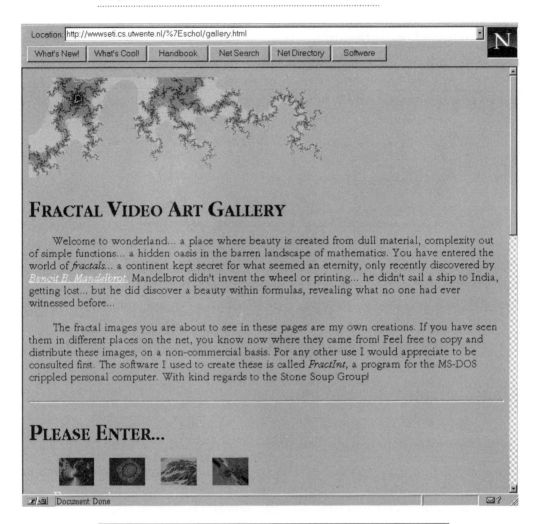

http://wwwseti.cs.utwente.nl/%7Eschol/gallery.html

Fractal Video Art Gallery features a wide-ranging collection of gorgeous images created with Fractint.

# FRACTALS MACHINE

http://phaethon.cti.gr/fractals.html

Check out this unique fractals zooming machine. It calculates the zooming parameters and uses the Fractint program (see below) to draw the fractal image. Zoom in or out on Mandelbrot and Julia sets.

# FRACTALS AND SCALE: A TUTORIAL

http://life.anu.edu.au:80/complex_systems/tutorial3.html

This brief and clear tutorial succinctly explains how and why Mandelbrot proposed the idea of a fractal as a way to cope with the problems of scale in the real world.

# FRACTALS AND THEIR APPLICATION TO GEOMETRY MODELS

http://www.fciencias.unam.mx/Graf/fractales/fract_l.html

Come here for some wonderful downloadable images including a region of the Mandelbrot set, and several superfaces modeled with fractals.

# THE FRACTINT PAR EXCHANGE

## http://www.heavanet.com/lkuhn/px

This page is for all the fans of Fractint (see the listing immediately below) and fractal creation. The main purpose of this page is for the trading of parameter files created by Fractint. Here you will find marvelous collections of unrendered fractals that are yours to play and experiment with.

# FRACTINT WORLD WIDE WEB PAGES

## http://spanky.triumf.ca/www/fractint/fractint.html

What is Fractint and what is it all about? This amazing software is a fractal generator for IBM PCs and compatibles as well as XWindow systems. It is by far the most versatile and extensive fractal program available at any price. And despite its excellence, it comes to you absolutely free. That's right. Fractint is freeware. The latest version of the software now features deep zooming. And it can be downloaded for free right here.

# GALLERY OF INTERACTIVE GEOMETRY

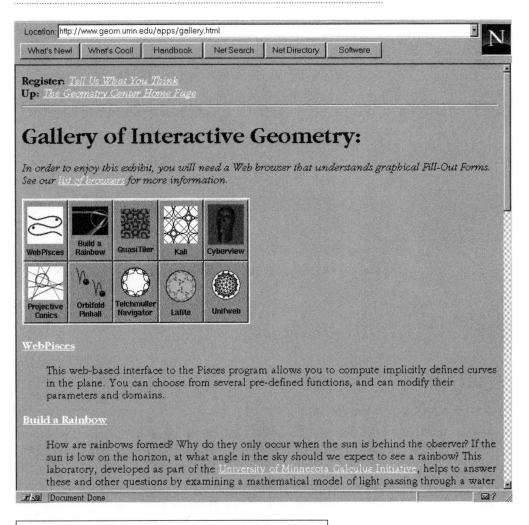

Location: http://www.geom.umn.edu/apps/gallery.html

What's New! | What's Cool! | Handbook | Net Search | Net Directory | Software

**Register:** *Tell Us What You Think*
**Up:** *The Geometry Center Home Page*

## Gallery of Interactive Geometry:

*In order to enjoy this exhibit, you will need a Web browser that understands graphical Fill-Out Forms. See our* list of browsers *for more information.*

WebPisces | Build a Rainbow | QuasiTiler | Kali | Cyberview

Projective Conics | Orbifold Pinball | Teichmuller Navigator | Lafite | Unifweb

### WebPisces

This web-based interface to the Pisces program allows you to compute implicitly defined curves in the plane. You can choose from several pre-defined functions, and can modify their parameters and domains.

### Build a Rainbow

How are rainbows formed? Why do they only occur when the sun is behind the observer? If the sun is low on the horizon, at what angle in the sky should we expect to see a rainbow? This laboratory, developed as part of the University of Minnesota Calculus Initiative, helps to answer these and other questions by examining a mathematical model of light passing through a water

Document Done

---

## http://www.geom.umn.edu/apps/gallery.html

What a wonderful collection of images, tools, and toys from the University of Minnesota. At the Gallery of Interactive Geometry you can:

❏ Work with WebPisces—This Web-based interface to the Pisces program allows you to compute implicitly defined curves in the

plane. You can choose from several predefined functions, and can modify their parameters and domains.

❑ Build a rainbow—How are rainbows formed? Why do they occur only when the sun is behind the observer? If the sun is low on the horizon, at what angle in the sky should we expect to see a rainbow? Begin to steer toward the answers to these and other questions by examining a mathematical model of light passing through a water droplet.

❑ Learn of Projective Conics—This discussion of Pascal's theorem in terms of projective geometry includes an interactive applications that lets you specify points on a conic and see how the theorem applies to them.

❑ Use QuasiTiler—Generate the famous Penrose tilings, or design your own nonperiodic tilings of the plane. In the process, you can select and visualize jplane cross-sections of a lattice in anywhere from 3 to 13 dimensions.

❑ Work with Kali and Kali-Jot—These programs are interactive editors for summetric patters of the plane, as seen in some of the woodcuts of M.C. Escher. It is also a fun way to learn about the seventeen crystallographic symmetry groups of the plane. But you need XMosaic.

❑ Have fun with Orbifold Pinball—Explore the effects of negatively curved space in this pinball-style game. The game board is not only curved, but also contains singularities that serve as "bumpers" off which the ball can bounce.

❑ Amaze and delight your friends with Teichmuller Navigator— Explore Teichmuller space, the space of all different angle geometries on a genus two surface. Moving through this space is accomplished by shifting the vertices of a tiling of the hyperbolic plane.

❑ Snag a free copy of Cyberview-X—No, this isn't pornographic. Cyberview-X is an interactive 3D viewer that works with any

HTML 2.0 compatible Web browser. You can pick an object out of the predefined library you'll find here, or you can learn about the OOGL format and define your own 3D objects.

❏ Use Unifweb—Discover and visualize families of Riemann surfaces with a specified group of symmetries. The presentation you choose for your symmetry group corresponds geometrically to a construction of the surface as a covering of a particular orbifold.

❏ Use Lafite—Work with any discrete symmetry group of the hyperbolic plane. Lafite will calculate the fundamental region and generators of the group you choose. The program than creates Escher-like patterns by replicating a motif through the action of the group.

# HAVING FUN WITH HYDRA

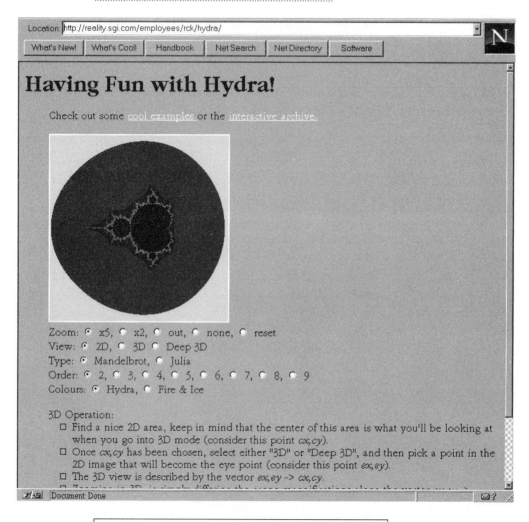

http://reality.sgi.com/employees/rck/hydra

Who was it said that chaos was where fun and science collide? This site is an example of the principle. Zoom in and out on a Mandelbrot or a Julia set, or both. You set the parameters and then you GO. This is all courtesy of Robert Keller (rck@sgi.com).

# ITERATED FUNCTION SYSTEMS PLAYGROUND

| http://www.cosy.sbg.ac.at/rec/ifs |
|---|

This great page includes an implementation of Tim Greer's Senile Depth First Search algorithm. Enter a negative iteration count to select this algorithm with "senility" equal to the absolute value of the number specified. Puzzled? Intrigued? See the January 1992 edition of the *IBM Technical Disclosure Bulletin* for a description of this algorithm.

# MANDELBROT EXHIBITION OF THE VIRTUAL MUSEUM OF COMPUTING

| http://www.comlab.ox.ac.uk/archive/other/museums/computing/ mandelbrot.html |
|---|

Come here for a marvelous collection of Mandelbrot set images. The site also provides interesting biographies of Benoit Mandelbrot and Gaston Julia. This space is maintained by Oxford University's Jonathan Bowen (Jonathan.Bowen@comlab.ox.ac.uk), to whom many thanks.

# MANDELBROT EXPLORER

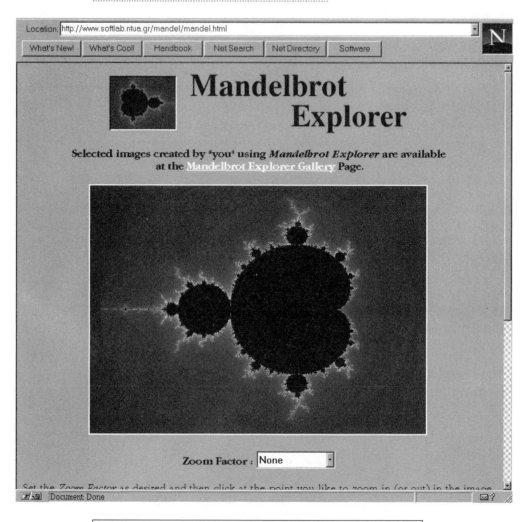

**http://www.softlab.ntua.gr/mandel/mandel.html**

Create new images, or zoom in or out on stored images, at this great site.

# MANDELBROT AND JULIA SET EXPLORER

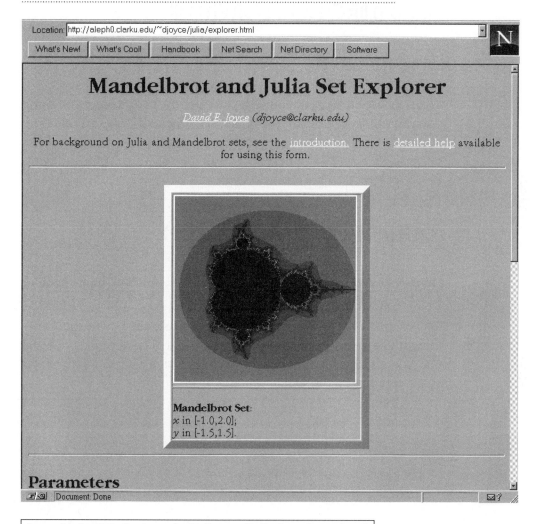

http://aleph0.clarku.edu/~djoyce/julia/explorer.html

Here is yet another site where you can enter parameters and create Mandelbrot and Julia sets.

# MANDELBROT SET: A JAVA-BASED MANDELBROT EXPLORER!!!!!!!

http://www.mindspring.com/~chroma/mandelbrot.html

You need a Sun Workstation and you need Sun's appletviewer. But if you've got those, prepare to be dazzled. Zoom in, out, around, and under 3D Mandelbrot images. A word of warning: the Solaris version of Netscape won't work here; it'll crash the software. You need Sun's appletviewer.

# MATHART.COM FRACTAL MUD

http://www.mathart.com/FractalMUD/FractalMUD_home.html

Oh yes. Just when you finally decided life was futile you found something fulfilling: an interactive exploration of a 3D deterministic fractal that you can help to shape as an evergrowing virtual world. Enter some parameters and create a new universe.

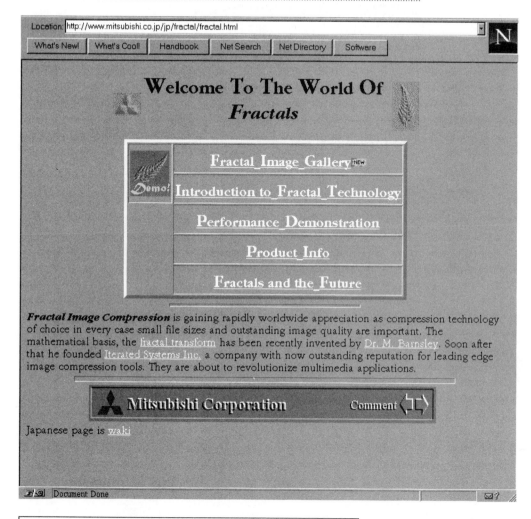

http://www.mitsubishi.co.jp/jp/fractal/fractal.html

Fractal image compression is rapidly gaining worldwide appreciation as the compression technology of choice in every case that small file sizes and outstanding image quality are important. Learn more about this technology, based on the fractal transform invented by Michael Barnsley, and how to leverage it at this informative Web site.

# NONLINEAR SCIENCE TODAY AND JOURNAL OF NONLINEAR SCIENCE

http://www.springer-ny.com/nst/

*Nonlinear Science Today* and *Journal of Nonlinear Science* provide forums for the dissemination of new results, methods, and ideas in nonlinear dynamics. Their goals are to facilitate the cross-fertilization and interaction between different areas of science, mathematics, and varied applied disciplines. Whereas the *Journal of Nonlinear Science* is a forum for scholarly articles, *Nonlinear Science Today* provides news and views for the dynamical community and an outlet for the discussion of current research at a generally accessible level, as well as more speculative ideas and opinions. Come to Springer-Verlag's Web page for these publications to get more information.

# NONLINEARITY AND COMPLEXITY HOME PAGE

http://www.cc.duth.gr/~mboudour/nonlin.html

The Nonlinearity and Complexity Home Page provides a rich cornucopia of links to conference proceedings, bibliographies, electronic journals, and more.

# CLIFFORD A. PICKOVER HOME PAGE

http://sprott.physics.wisc.edu/pickover/home.html

Access this home page for one of our favorite science writers and novelists, the author of many splendid books and articles on chaos theory and fractal science. Pickover is currently a research staff member at the IBM T.J. Watson Research Center, where he has received eleven invention achievement awards, a research division award, and four external honor awards. He is also a consultant for WNET on science education projects and a regular columnist for *Discover* magazine.

# SOFTWARE FOR NONLINEAR DYNAMICAL SYSTEMS

http://www.physik.th_darmstadt.de/nlp/distribution.html

What free software this site will yield! Three items worth talking about:

❐ Dimension—calculates the Grassberger-Procaccia, the Termonia-Alexandrowicz, and the average pointwise dimension of data sets;

❐ Lyapunov Exponents—calculates the spectrum of Lyapunov exponents via a linear approximation of the flow constructed from an experimental time series

❐ Time Series Analyzer—a graphical environment for nonlinear time series and chaotic behavior.

# SPANKY FRACTAL DATABASE

http://spanky.triumf.ca/

The Spanky Fractal Database is a collection of fractals and fractal-related material. Most of the software was gathered from various ftp sites on the Internet and it is generally freeware or shareware. Here you have fractal images, fractal programs, fractal tutorials, interactive fractal explorers, and much more.

# TEST FRACTAL GENERATOR

http://mendel.berkeley.edu/~seidel/testfrac.html

Enter some values into the convenient form, and then the computer will iterate them and generate a GIF (Graphics Interchange Format) image. Couldn't be easier.

# VRML FRACTALS

http://kirk.usafa.af.mil/%7Ebaird/vrml

Use Webspace to view these fractals. You can download the Webspace software at this site, in fact, and then use it to view a 3D VRML fractal tree and a 3D Cantor set. You can also use Worldview, but note that Worldview cannot display full fractals but rather only small pseudofractals that appear identical to real fractals at a distance but have less detail close up. The number of leaves on a pseudofractal tree is an exponential function of the size of the the VRML file, so even pseudofractals allow fairly interesting models. For Worldview you get an image of a great 3D pseudofractal mountain. You also get the C source code for the mountain (incorporating equations derived from Mathematica code).

# WATERLOO-MONTREAL-VERONA FRACTAL RESEARCH INITIATIVE

http://links.uwaterloo.ca/

The Waterloo-Montreal-Verona Fractal Research Initiative is a collaborative effort designed to further the theoretical understanding of fractal mathematics and its application to signal processing. Toward this end the personnel of the initiative have assembled a unique and useful set of resources on the Web.

# WHAT IS A FRACTAL?

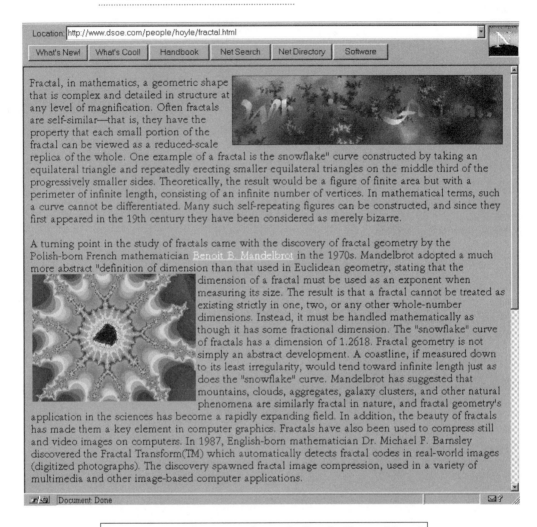

Location: http://www.dsoe.com/people/hoyle/fractal.html

| What's New! | What's Cool! | Handbook | Net Search | Net Directory | Software |

Fractal, in mathematics, a geometric shape that is complex and detailed in structure at any level of magnification. Often fractals are self-similar—that is, they have the property that each small portion of the fractal can be viewed as a reduced-scale replica of the whole. One example of a fractal is the snowflake" curve constructed by taking an equilateral triangle and repeatedly erecting smaller equilateral triangles on the middle third of the progressively smaller sides. Theoretically, the result would be a figure of finite area but with a perimeter of infinite length, consisting of an infinite number of vertices. In mathematical terms, such a curve cannot be differentiated. Many such self-repeating figures can be constructed, and since they first appeared in the 19th century they have been considered as merely bizarre.

A turning point in the study of fractals came with the discovery of fractal geometry by the Polish-born French mathematician Benoit B. Mandelbrot in the 1970s. Mandelbrot adopted a much more abstract "definition of dimension than that used in Euclidean geometry, stating that the dimension of a fractal must be used as an exponent when measuring its size. The result is that a fractal cannot be treated as existing strictly in one, two, or any other whole-number dimensions. Instead, it must be handled mathematically as though it has some fractional dimension. The "snowflake" curve of fractals has a dimension of 1.2618. Fractal geometry is not simply an abstract development. A coastline, if measured down to its least irregularity, would tend toward infinite length just as does the "snowflake" curve. Mandelbrot has suggested that mountains, clouds, aggregates, galaxy clusters, and other natural phenomena are similarly fractal in nature, and fractal geometry's application in the sciences has become a rapidly expanding field. In addition, the beauty of fractals has made them a key element in computer graphics. Fractals have also been used to compress still and video images on computers. In 1987, English-born mathematician Dr. Michael F. Barnsley discovered the Fractal Transform(TM) which automatically detects fractal codes in real-world images (digitized photographs). The discovery spawned fractal image compression, used in a variety of multimedia and other image-based computer applications.

Document: Done

## http://www.dsoe.com/people/hoyle/fractal.html

What is "What Is a Fractal?" Well, a wonderfully executed HTML tutorial, of course. This is a splendid place to start your journey into chaos.

# CHEMISTRY

Analytical Chemistry Databases and Publications
Basic Liquid Chromatography: Hypertextbook
Biological Mass Spectrometry Software
The Boron-Mediated Aldol Reaction
Charged Particle Optics Software
Chemical Calculations
Chemistry Hypermedia Project at Virginia Tech
Chemistry Principles Hypertext
Chemistry Teaching Resources
Chemist's Art Gallery
The Combustion Chemistry Laboratory
Computational Chemistry for Chemistry Educators: A
    Tutorial
Computational Chemistry and Organic Synthesis: A Tutorial
Computers in Chemistry Education
The Edison Project for Communicating Chemistry
EMSL Gaussian Basis Set Generator
HPLC Troubleshooter
Hydrogen Bonding: Atlas of Side-Chain and Main-Chain
    Bonding

Indiana University Biogeochemical Laboratories
Inorganic Elemental Analysis at Los Alamos
Internet Chemistry Resources
Laboratory for Analytical Chemistry
MIME Types for Chemistry
MOLSCAT, version 14
Murray's Mass Spectrometry on the Internet
The Multimedia Education Library (MEL)
NMR Analysis Using Hyperactive Molecules: An Illustrated
    Tutorial
Organic Chemistry Resources Project
Periodic Table: Chemicool
Periodic Table: Illinois Institute of Technology
Periodic Table: Illinois State Geological Survey
Polymer Chemistry Hypertext
Sherpa: Your Guide to the Peaks
TIMS: Tool for the Interpretation of Mass Spectra
World Wide Web Hub for Chemistry

# ANALYTICAL CHEMISTRY DATABASES AND PUBLICATIONS

http://www.nist.gov/srd/analy.htm

Here you will find a comprehensive set of easy-to-use databases meant to help the analytical chemist identify unknown materials, and in many cases, once identified, avoid the need to recharacterize a substance. The data have been fully evaluated using a variety of techniques. When appropriate, duplicate measurements have been included for completeness. The databases are updated and expanded on a regular basis and have powerful, fast search engines. The databases cover National Institute of Standards and Technology (NIST), Environmental Protection Agency (EPA), and National Institutes of Health (NIH) files for mass spectra, gas-phase infrared, x-ray photoelectron spectroscopy, surface structure, electron diffraction, and crystal data.

# BASIC LIQUID CHROMATOGRAPHY: HYPERTEXTBOOK

http://128.173.180.166/my_home/book/content.html

Come here for an absolutely terrific hypertext tutorial on basic LC techniques. The tutorial includes complete coverage of:

❒ general theory—including kinetics (band broadening) and thermodynamics of adsorption from solutions;

❒ adsorbents—structure, surface chemistry, bonded phases, and major parameters of packing materials;

❒ reversed-phase high-performance liquid chromatography (HPLC) —including retention mechanism and influence of the bonded phase type and density, mobile phase composition, and other parameters;

❏ instrumentation—pumps, injectors, connectors, and more;

❏ detectors—RI, UV, diode array, fluorescence, LC/MS, ELS and other HPLC detectors.

# BIOLOGICAL MASS SPECTROMETRY SOFTWARE

> http://mac-mann6.embl-heidelberg.de/MassSpec/Software.html

Download free software including:

❏ PeptideSearch (Mac) for protein identification in sequence databases by searching with mass spectrometric peptide data;

❏ MacBioSpec (Mac), a versatile software package for detailed protein and peptide analysis;

❏ general protein mass analysis software (PC/Windows) for interpretation of peptide and protein mass data;

❏ protein analysis worksheet software (PC & Mac) for interpretation of protein and peptide mass spectrometry data;

❏ several more Mac and PC tools for small molecule analysis.

# THE BORON-MEDIATED ALDOL REACTION

> http://www.ch.cam.ac.uk/MMRG/aldol1.html

These Web pages describe an investigation of the boron-mediated aldol reaction undertaken by Dr. Ian Paterson and Dr. Jonathan Goodman at the University of Cambridge, and by Professor Cesare Gennari and Professor Anna Bernardi at the University of Milan. The boron-mediated aldol is a useful reaction in organic synthesis, because it can form two new chiral centers while forming a carbon-carbon bond.

## CHARGED PARTICLE OPTICS SOFTWARE

http://wwwdo.tn.tudelft.nl/bbs/cposis.htm

Access PC and Mac software created to design and simulate properties of instruments utilizing electric and magnetic fields to guide charged particles: electron is, positrons, and ions.

## CHEMICAL CALCULATIONS

http://pasd2.paschools.pa.sk.ca/chemical/

Check out this great set of experimental chemical calculators. One calculates molar mass, the other percentage composition.

## CHEMISTRY HYPERMEDIA PROJECT AT VIRGINIA TECH

http://www.chem.vt.edu/chem-ed/vt-chem-ed.html

The Chemistry Hypermedia Project is developing tutorials that provide supplemental educational resources for undergraduate chemistry students. The hypermedia documents contain hyperlinks to remedial material that describes fundamental chemical principles. Topics covered include analytical chemistry and solid-state chemistry.

# CHEMISTRY PRINCIPLES HYPERTEXT

http://neon.unm.edu/Courses/General/Lecture/Principles.html

Here is an electronic edition of an excellent textbook on chemical principles by Don McLaughlin of the University of New Mexico. *Chemistry Principles* is in PostScript format for viewing with Ghostview, which is commonly installed with most Mosaic clients.

# CHEMISTRY TEACHING RESOURCES

http://www.anacheim.umu.se/eks/pointers.htm

This site provides links to dozens of resources including hypertexts, curriculum guides, graphics and visualizations, on-line journals, and downloadable software for a range of platforms.

# CHEMIST'S ART GALLERY

http://www.csc.fi/lul/chem/graphics.html

The Chemist's Art Gallery contains spectacular visualizations and animations. These include animations of small molecule diffusion in polymers, animations of the protein *cellobiohydrolase I*, visualization of volumes of chromosomes and viruses based on electron microscopy tomography, visualization of micelles, and visualization of the dynamics of the spreading of small droplets of chainlike molecules on surfaces.

# THE COMBUSTION CHEMISTRY LABORATORY

http://mephisto.ca.sandia.ga/

The Combustion Chemistry Laboratory is a part of the Combustion Research Facility at Sandia National Laboratories. This is a group of keneticists studying elementary chemical reactions important in combustion processes. For the last several years the staff of the lab has been studying the chemistry of NHx species. This chemistry is important in the formation and destruction of NOx, an important class of pollutants. The site includes on-line hypertext presentations on such topics as quantum Monte Carlo evaluation of chemical reaction rate coefficients, protodediazoniation of an aryldiazonium atom, the reaction of NH2 with oxygen, and more.

# COMPUTATIONAL CHEMISTRY FOR CHEMISTRY EDUCATORS: A TUTORIAL

http://www.mcnc.org/HTML/ITD/ncsc/ccsyllabus.html

The goal of this hypertext tutorial is to introduce chemistry educators to computational techniques in quantum chemistry. The course is intended as an enrichment activity, designed to help educators understand the use of high-performance computing tools and their role in chemical research. The tutorial comes courtesy of the North Carolina Supercomputing Center (NCSC). Using commercial third-party software (Gaussian94 and MOPAC) resident on the North Carolina Supercomputing Center's Cray Y-MP supercomputer, educators can investigate a number of chemical properties, such as single-point energies, geometry optimizations, frequency calculations, potential energy surfaces, and thermochemical properties.

# COMPUTATIONAL CHEMISTRY AND ORGANIC SYNTHESIS: A TUTORIAL

http://www.caos.kun.n/~borkent/compcourse/comp.html

This hypertext tutorial is meant to introduce computational chemistry to synthetic-organic chemists who would like to understand why they got the product they got and not (always) the compound they wanted. The tutorial emphasizes the Transition State (TS): how to construct it, characterize it, and compare it to other alternatives. The tutorial begins with conformational changes and ends with a study of "real" reactions.

# COMPUTERS IN CHEMISTRY EDUCATION

http://www.liv.ac.uk/ctichem.html

This Web site is maintained by the CTI Center for Chemistry, based in the Chemistry Department of the University of Liverpool. The Center works to encourage the use of learning technologies in higher education. Here you will find hypertext curriculum guides, textbooks, educational software, course outlines, and more.

# THE EDISON PROJECT FOR COMMUNICATING CHEMISTRY

http://www.columbia.edu/cu/chemistry/Edison.html

The Edison Project at Columbia University provides several downloadable multimedia tutorial modules primarily for the Macintosh. Topics include substitution/elimination reactions, conformations of butane, molecular representations, VSEPR, glucose mutarotation, periodic motion, and Lewis Dot structures.

## EMSL GAUSSIAN BASIS SET GENERATOR

http://www.emsl.pnl.gov:2080/forms/basisform.html

Enter the elements you wish to calculate and push the button. Your output can, if you wish, optimize general contractions and/or show supported elements.

## HPLC TROUBLESHOOTER

http://helium.fct.unl.pt/gof/hplcts.html

The doctor is in. Just point to where it hurts. No peaks? Very small peaks? No flow? Low pressure? High pressure? Variable retention? Change on separation? Split peaks? Tailing? Fronting? Rounded peaks? Base line drift? Base line noise? Broad peaks? Changes in peak height? Negative peaks? Ghost peaks? Point to the problem, click, and get the cure.

## HYDROGEN BONDING: ATLAS OF SIDE-CHAIN AND MAIN-CHAIN BONDING

http://bsmcha1.biochem.ucl.ac.uk:80/~mcdonald/atlas

This hypertext document is a graphical summary of hydrogen bonding in a data set of high-resolution protein structures. It shows the distributions of the frequencies and geometries of hydrogen bonds formed by main-chain and side-chain donors and acceptors. The document is designed to serve as a reference on hydrogen bonding patterns of the standard hydrogen-bonding groups for molecular modelers and crystallographers who would like to know whether the hydrogen bonding in the model is normative. With regard to side-chain bonding, the document addresses Arginine, Asparagine, Aspartate, Cysteine, Cystine, Glutamate, Glutamine, Histidine, Lysine, Methionine, Serine, Threonine, Tryptophan, and Tyrosine.

# INDIANA UNIVERSITY BIOGEOCHEMICAL LABORATORIES

http://silver.ucs.indiana.edu/~isotopes/home.html

The Biogeochemical Labs at Indiana University are the birthplace of isotope ratio monitoring gas chromatography/combustion/mass spectrometry. Here the technique continues to be refined in terms of both hardware and software, and to be applied to different gases. This site includes details on the latest work at the lab, an overview of the GC/C/MS technique, a bibliography of articles describing and/or using the GC/C/MS technique, documentation and updates for the ISODAT software system, and more.

# INORGANIC ELEMENTAL ANALYSIS AT LOS ALAMOS

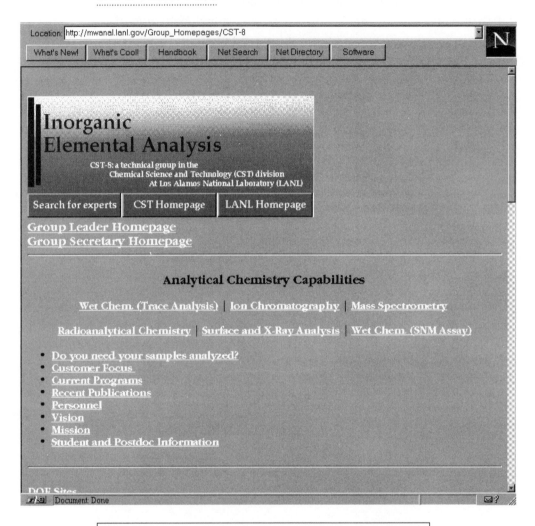

http://mwanal.lanl.gov/Group_Homepages/CST-8

Access the Web home of CST-8, a technical group in the Chemical Science and Technology (CST) division at Los Alamos National Laboratory. The mission of CST-8 is to provide inorganic elemental analysis of both radioactive and nonradioactive materials for the National

Laboratory and for the nation. The primary focus of the group is measurement of percent and higher levels of constituents using traditional and state of the art instrumental analytical methods, provide analytical services, develop and improve analytical methods, and perform R&D in support of new methods and techniques. The analytical chemistry capabilities of the lab include wet chemistry (trace analysis), ion chromatography, mass spectrometry, radioanalytical chemistry, surface and x-ray analysis, and SNM assays.

# INTERNET CHEMISTRY RESOURCES

| http://www.rpi.edu/dept/chem/cheminfo/chemres.html |

The Internet Chemistry Resources list is a highly selective list of resources related to chemistry and associated fields. Resources are organized according either to type of Internet service (gopher, ftp, etc.) or subject (e.g., teaching resources). Come here for access to:

❏ catalogs and publishers

❏ databases and data collections

❏ document delivery

❏ e-mail servers and LISTSERVS

❏ FTP resources

❏ Gophers

❏ on-line search services

❏ periodicals and conference proceedings

❏ other documents

❏ software archives

❒ teaching resources

❒ corporate Web resources.

Additionally, you may search the entirety of the Chemistry Resources list using the built-in *glimpse* search engine.

# LABORATORY FOR ANALYTICAL CHEMISTRY

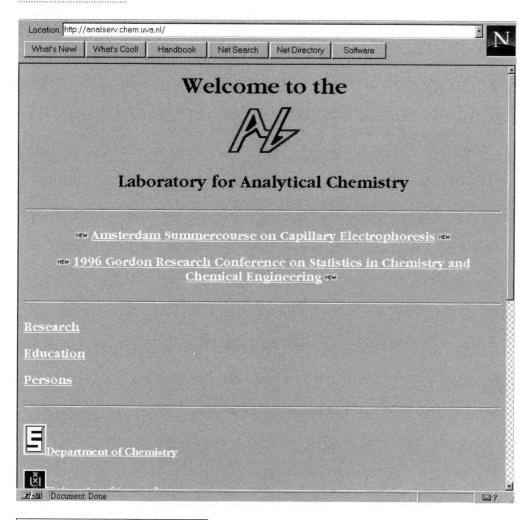

**http://analserv.chem.uva.nl**

Access the research resources of the Laboratory for Analytical Chemistry at the University of Amsterdam. The site provides details on the work of the Lab's research groups involved with separation methods as well as process analysis and chemometrics.

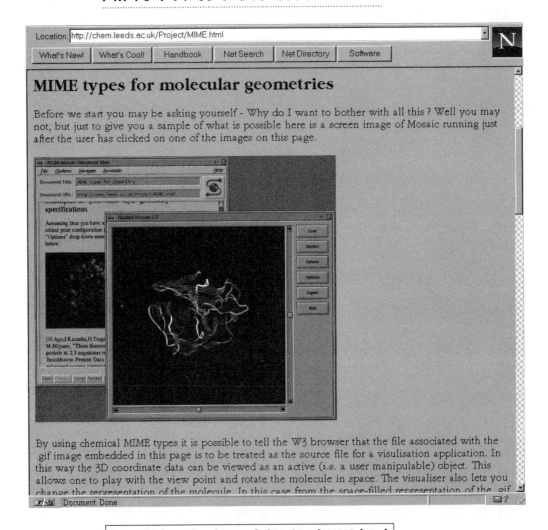

## MIME types for molecular geometries

Before we start you may be asking yourself - Why do I want to bother with all this? Well you may not, but just to give you a sample of what is possible here is a screen image of Mosaic running just after the user has clicked on one of the images on this page.

By using chemical MIME types it is possible to tell the W3 browser that the file associated with the .gif image embedded in this page is to be treated as the source file for a visulisation application. In this way the 3D coordinate data can be viewed as an active (i.e. a user manipulable) object. This allows one to play with the view point and rotate the molecule in space. The visualiser also lets you change the representation of the molecule. In this case from the space-filled representation of the gif

**http://chem.leeds.ac.uk/Project/MIME.html**

By using chemical MIME types it is possible to tell a W3 browser that a file associated with a given .gif image is to be treated as the source for a visualization application. In this way the 3D coordinate data can be viewed as an active (i.e., a user manipulable) object. This allows one to play with the viewpoint and rotate the molecule in space. The visualizer also lets you change the representation of the molecule.

A common format for describing the structure of a large biological molecule is the protein database of "pdb" file. Large databases such as the "Molecules-R-Us" National Institutes of Health database, the Brookhaven Protein Data Bank, or the UK Chemistry Database Service can be easily accessed through the Web and its gateways. Given the information and tools contained at this Web site, it is a simple matter to reconfigure a Web browser to view these files as active 3D objects. (To further help you along, the site provides hyperlinks to all three databases named above.)

# MOLSCAT, VERSION 14

http://molscat.giss.nasa.gov/MOLSCAT

The MOLSCAT software available here is a code for quantum mechanical (coupled channel) solution of the nonreactive molecular scattering problem. The software is implemented for various types of collision partners. In addition to the essentially exact close coupling method, several approximate methods, including coupled states and infinite order sudden approximations, are possible. The software code has been written in near standard FORTRAN 77 and has been ported to a large number of platforms. The source code at this site is currently running on an IBM RS/6000 and also on IBM and compatible mainframes. It should also work on most other machines, although word has it there are some bugs to contend with when one runs it on a Cray.

# MURRAY'S MASS SPECTROMETRY ON THE INTERNET

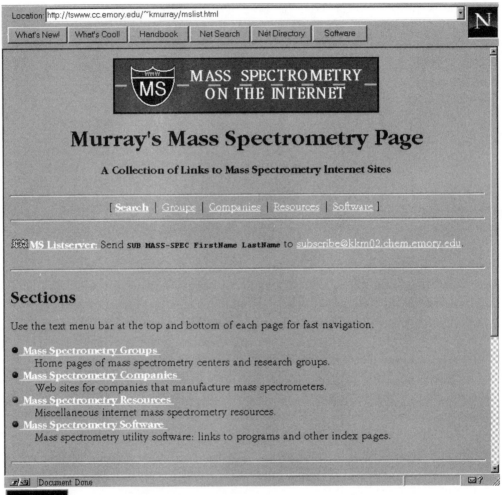

http://tswww.cc.emory.edu/~kmurray/mslist.html

Come here for hundreds of links to mass spectrometry centers and research groups, manufacturers of mass spectrometers, and mass spectrometry utility software.

# THE MULTIMEDIA EDUCATION LIBRARY (MEL)

http://www.engin.umich.edu/labs/mel/index.html

The Multimedia Education Library (MEL) creates computer-based modules for the advancement of chemical engineering understanding. Here is a list of what you can download for free:

❑ Ford-Wixom Phosphate Coating System Plant Tour—This module is an interactive tour of the Ford Pre-Paint/Phosphate Coating Plant at Wixom, Michigan. Students have to create a flowsheet of the system as well as a complete mass balance. The ultimate goal of the project is to suggest ways to minimize waste production at the Ford-Wixom Plant.

❑ Process Parameters—Here is an interactive investigation of the concepts and applications of pressure and temperature. It includes questions that relate these concepts to the user's everyday experiences, as well as images of industrial pressure and temperature measuring equipment.

❑ Multiphase Systems—This interactive module explores single and multicomponent systems. It covers topics such as phase diagrams, phase equilibria, and phase transitions. Separate sections include examples of industrial applications of these concepts.

❑ The Pump Experience—An interactive tutorial that illustrates various types of pumps, characteristic curves, and liquid flow through different pump systems (ideal as a supplemental aid in fluid mechanics course work).

❑ Chemical Engineering Equipment—A visual overview of the various types of equipment used in chemical engineering today, including their specific applications, design, advantages, and disadvantages.

❑ Receptors—These are links of the interaction of cell receptors and ligands with chemical engineering kinetics concepts.

❐ Adsorption Separation—This module focuses on the applications of affinity chromatography to biotechnology. The general equations for adsorption separation are derived and a chromatography column simulator can be used to explore the effect of various parameters on column performance.

## NMR ANALYSIS USING HYPERACTIVE MOLECULES: AN ILLUSTRATED TUTORIAL

http://www.ch.ic.ac.uk/rzepa/mjce/

From Henry S. Rzepa and Christopher Leach at the Department of Chemistry, Imperial College, London, comes a splendid Web tutorial on NMR analysis with hyperactive molecules.

## ORGANIC CHEMISTRY RESOURCES PROJECT

http://yip5.chem.wfu.edu/yip/organic/org-home.html

The best thing about this site is some good software available for free downloading. The software is available in both Mac and Windows versions, and is designed to demonstrate derivation of the Rydberg constant, electron distribution of the 2s orbital, the shape of the $4dx2-y2$ wave function, the size of the $3dz2$ orbital, and more. There is also a linear regression graphing pad, which you can either download or use on-line.

# PERIODIC TABLE: CHEMICOOL

http://the-tech.mit.edu/~davhsu/chemicool.html

This graphics periodic table implementation provides for each element general information, states, energies, oxidation and electrons, appearance and characteristics, reactions, other forms, radius, conductivity and abundance, as well as links to *Encyclopedia Britannica* articles. There's also a less graphic option in case you are on a slow link. In addition to pure periodic data, this site also offers graphs of atomic weight, atomic radius, and ionization energy vs. atomic number.

# PERIODIC TABLE: ILLINOIS INSTITUTE OF TECHNOLOGY

http://www.csrri.iit.edu/periodic-table.html

A click on an element name button in this forms-based periodic table will give you the x-ray properties of that element. If you give an energy value in a box at the top of the table you also get x-ray cross sections at that energy. The source code used to calculate x-ray cross sections is also available.

# PERIODIC TABLE: ILLINOIS STATE GEOLOGICAL SURVEY

http://steele.isgs.uiuc.edu/pt/

This link brings you to both a table-based and a text-only periodic table implementation.

# POLYMER CHEMISTRY HYPERTEXT

> http://www.umr.edu/~wlf/

This hypertext tutorial and textbook is a valuable tool for learning about adhesion (Bushe-Cashin-Debye equation, chain expansion, and chain transfer), coatings (including the coefficient of curvilinear diffusion), mechanical properties, molecular weight, polymer solutions, time-temperature superposition, and viscosity. The site includes a great hyperlinked index of polymer movies available on the Web.

# SHERPA: YOUR GUIDE TO THE PEAKS

> http://128.95.12.16/Sherpa.html

Sherpa is a Macintosh-based expert system for LC/MS and MS/MS analysis. Sherpa is designed to be a robust, easy to use aide-de-camp in the correlation and interpretation of LC/MS and MS/MS spectra to a known protein sequence. Sherpa has an easy-to-use graphical interface and offers simultaneous analysis against multiple protein sequences. It also enables the comparison of two LC/MS files, offers posttranslational modification searches for glycosylation and phosphorylation, includes calculation of enzymatid digest peptide masses, and lets the user define custom residues and termini.

# TIMS: TOOL FOR THE INTERPRETATION OF MASS SPECTRA

> http://beelzebub.ethz.ch/TBres.html

TIMS is a simple Mac tool with a graphic user interface that helps the interpretation of mass spectra. After the user has provided an experimental mass spectrum and a possible structure (connection table) TIMS calculates all possible fragments obtained removing one or two bonds. Isotope signals are automatically generated. Download a free copy of TIMS right here.

# WORLD WIDE WEB HUB FOR CHEMISTRY

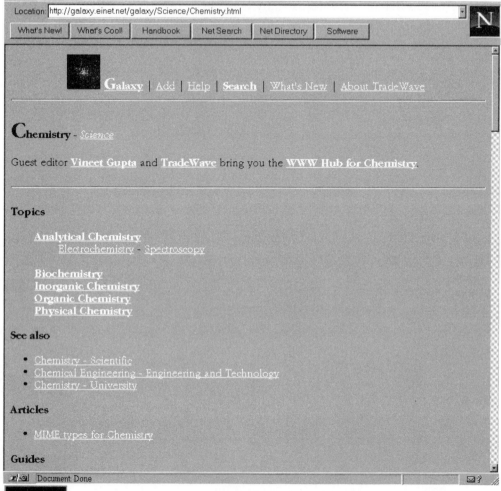

**http://galaxy.einet.net/galaxy/Science/Chemistry.html**

I've saved the best chemistry site for last. If you load just one chemistry site into the hotlist on your browser, this should be it. From this metasite you can access hundreds of great Web documents and archives related to analytical chemistry (electrochemistry and spectroscopy), biochemistry, inorganic chemistry, organic chemistry, and physical chemistry. Here you have on-line journals, downloadable software, data archives, laboratory and academic department home pages, and much, much more.

# FUN STUFF

Hot Air: Home Page of the **Annals of Improbable Research**
Mad Science
Science Jokes Archive

Science Made Stupid
Studmuffins of Science

# HOT AIR: HOME PAGE OF THE ANNALS OF IMPROBABLE RESEARCH

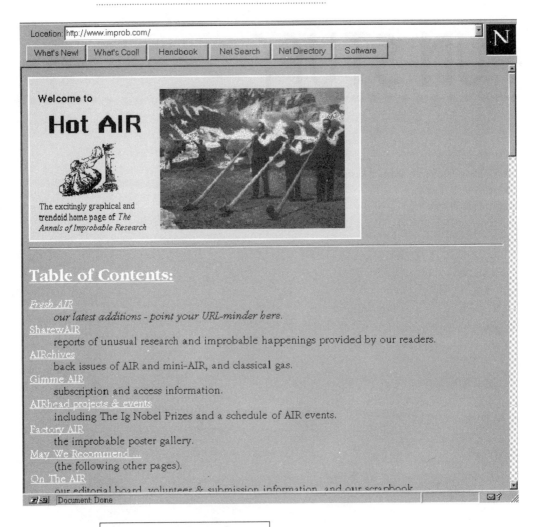

**http://www.improb.com**

Here at the home page for this remarkable publication you will find lots of fun stuff including downloadable posters (suitable for framing or birdcages). What else? How about cool, fun software and tutorials including *A Periodic Table of the Presidents*, *Fun with Grapes*, *An Ode to the World's First Rubber Czech*, and *Plans for Things That Go Boom*.

# MAD SCIENCE

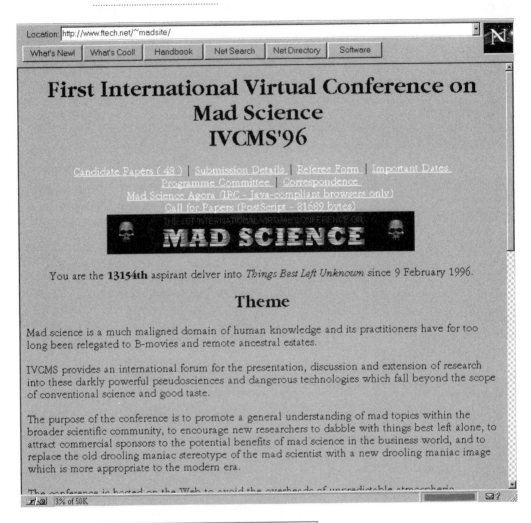

## First International Virtual Conference on Mad Science IVCMS'96

Candidate Papers ( 48 ) | Submission Details | Referee Form | Important Dates
Programme Committee | Correspondence
Mad Science Agora (IRC - Java-compliant browsers only)
Call for Papers (PostScript - 81689 bytes)

THE 1ST INTERNATIONAL VIRTUAL CONFERENCE ON
**MAD SCIENCE**

You are the **13154th** aspirant delver into *Things Best Left Unknown* since 9 February 1996.

### Theme

Mad science is a much maligned domain of human knowledge and its practitioners have for too long been relegated to B-movies and remote ancestral estates.

IVCMS provides an international forum for the presentation, discussion and extension of research into these darkly powerful pseudosciences and dangerous technologies which fall beyond the scope of conventional science and good taste.

The purpose of the conference is to promote a general understanding of mad topics within the broader scientific community, to encourage new researchers to dabble with things best left alone, to attract commercial sponsors to the potential benefits of mad science in the business world, and to replace the old drooling maniac stereotype of the mad scientist with a new drooling maniac image which is more appropriate to the modern era.

**http://www.ftech.net/~madsite/**

"Mad science is a much maligned domain of human knowledge and its practitioners have for too long been relegated to B-movies and remote ancestral estates." So writes the Webmaster who maintains the Mad Science Web site.

If you would care to delve into things best left unknown, then plant your tongue firmly in your cheek and visit Mad Science, the international forum for the presentation, discussion, and extension of research into darkly powerful pseudosciences and dangerous technologies that fall beyond the scope of conventional science and even good taste.

The forum is in fact an ongoing cyber conference designed "to promote a general understanding of mad topics with the broader scientific community, to encourage new research to dabble with things best left alone, to attract commercial sponsors to the potential benefits of mad science in the business world, and to replace the old drooling maniac stereotype of the mad scientist with a new drooling maniac image which is more appropriate to the modern era." The conference is held on the Web, rather than at a site outside the virtual world, in order to avoid high overhead, unpredictable atmospheric conditions, and revolting peasants.

Topics of specific interest include, but are not limited to:

❏ Creating life to satisfy egocentric motives

❏ Unleashing entities beyond human control and comprehension

❏ Tampering with life-sustaining forces of the universe

❏ Exceeding the limitations of the human body via grotesque metamorphoses

❏ New applications for old technologies (alchemy, necromancy, etc.)

❏ Ill-advised collaboration with alien and/or supernatural intelligences

❏ Lifelong devotion to researching the pointless and inane

❏ Callous disregard for human experimental subjects

❏ The art of exacting bizarre revenge on contemptuous and derisive peers.

# SCIENCE JOKES ARCHIVE

http://www.princeton.edu/~pemayer/ScienceJokes.html

Take your pick. We've got dozens of math jokes, physics jokes, chemistry jokes, biology jokes, and more, including a complete collection of "The mathematician, the physicist, and the engineer" jokes.

# SCIENCE MADE STUPID

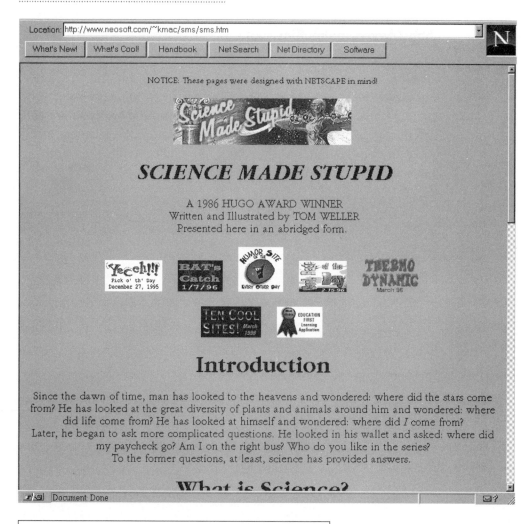

This is the 1986 Hugo Award–winning book written and illustrated by Tom Weller, here presented in an abridged but still very funny form.

# STUDMUFFINS OF SCIENCE

http://www.studmuffins.com

Check in and see who's been nominated as a studmuffin of science, and then vote your choice. Or make your own nominations. "If you know a hunky science prof or cute guy in the lab (or if you are cute and hunky), LET US KNOW" reads the invitation. I'd let them know about myself but I'm sure I've been nominated more than a dozen times already!

# GENERAL SCIENCE

# BAD SCIENCE

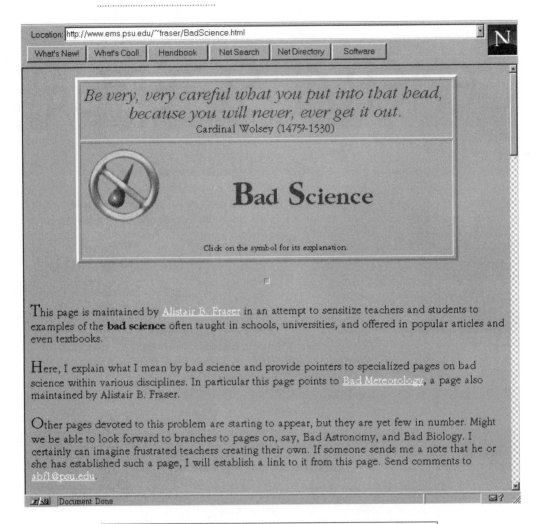

http://www.ems.psu.edu/~fraser/BadScience.html

This page is maintained by Alistair B. Fraser in an attempt to sensitize teachers and students to examples of the bad science often taught in schools and universities, and offered in popular articles and even some textbooks. Fraser explains what he means by bad science and provides a list of pointers to specialized pages on bad science within various disciplines, including bad chemistry and bad meteorology.

Bad science is defined within the context of this page as "well understood phenomena which are persistently presented incorrectly by teachers and writers, presumably because they either do not know any better or because they don't really care enough to get it correct."

# DR. INTERNET'S SCIENCE RESOURCES

http://ipl.sils.umich.edu/youth/DrInternet/

This Web page helps children explore science and math through fun, interactive exercises. As the Webmaster explains, "Dr. Internet will help you to find stuff that can help with your homework or your science project of that is just cool!" This is a remarkably well done, carefully selected set of science links for elementary school children.

# EARTH AND SKY RADIO SERIES

http://www.earthsky.com/

I'm sure you are familiar with the radio program "Earth and Sky." If you are not, you should be. The show is heard by millions of listeners on over 500 commercial and public stations throughout the United States, Canada, and the South Pacific. It is also aired on a variety of international networks, including Armed Forces Radio, the World Radio Network, and Voice of America.

Each day, Deborah Byrd and Joel Block discuss popular science subjects that affect our everyday lives. Each program is a snapshot of current scientific knowledge, progress, and questions, presented in language we all can understand and appreciate.

Come here for news of upcoming programs, and other information about "Earth and Sky."

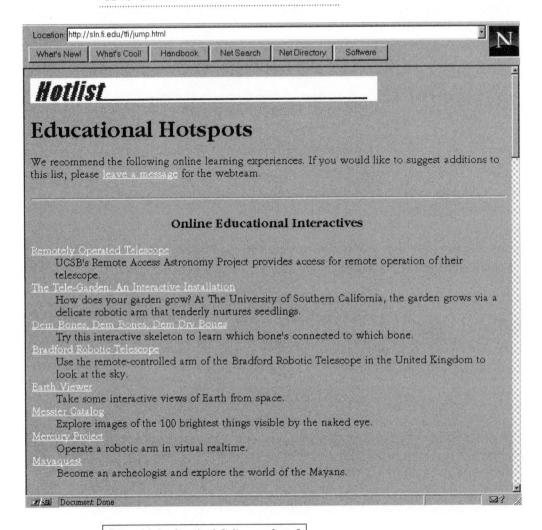

Location: http://sln.fi.edu/tfi/jump.html | N

| What's New! | What's Cool! | Handbook | Net Search | Net Directory | Software |

## *Hotlist*

## Educational Hotspots

We recommend the following online learning experiences. If you would like to suggest additions to this list, please leave a message for the webteam.

### Online Educational Interactives

Remotely Operated Telescope
UCSB's Remote Access Astronomy Project provides access for remote operation of their telescope.
The Tele-Garden: An Interactive Installation
How does your garden grow? At The University of Southern California, the garden grows via a delicate robotic arm that tenderly nurtures seedlings.
Dem Bones, Dem Bones, Dem Dry Bones
Try this interactive skeleton to learn which bone's connected to which bone.
Bradford Robotic Telescope
Use the remote-controlled arm of the Bradford Robotic Telescope in the United Kingdom to look at the sky.
Earth Viewer
Take some interactive views of Earth from space.
Messier Catalog
Explore images of the 100 brightest things visible by the naked eye.
Mercury Project
Operate a robotic arm in virtual realtime.
Mayaquest
Become an archeologist and explore the world of the Mayans.

Document: Done

## http://sln.fi.edu/tfi/jump.html

**META SITE**

Here is a list of links to interactive science learning experience on the Web, including:

❑ UCSB's Remote Access Astronomy Project, which provides access to the operation of their telescope via the Web;

- The Tele-Garden, whereby you can control a delicate robotic arm through the Web that tenderly nurtures seedlings at the University of Southern California;

- Dem Bones, Dem Bones, Dem Dry Bones—try this interactive skeleton to learn which bone's connected to which bone;

- Earth Viewer—enjoy some interactive views of Earth from space;

- Mercury Project—operate a robotic arm in virtual real-time;

- MayaQuest—become an archaeologist and explore the world of the Mayans.

# FACTOIDS OF SCIENCE

http://www.gene.com/ae/wn/Factoids/factoids.html

Here are facts with which you can amuse, inspire, and challenge students and colleagues. A few samples:

- Global Cooling?—The composite global temperature in the lower atmosphere was below average for the second consecutive month, and a record low temperature in the stratosphere was recorded, in January 1996. This could be caused by the greenhouse effect and/or ozone depletion.

- ENICAC, the first electronic computer, appeared 50 years ago in February 1946. The original ENIAC was about 80 feet long, weighted 30 tons, and had 17,000 tubes. By comparison, a desktop computer today can store a million times more information than an ENIAC, and run 50,000 times faster.

- Amazonian Plot—A plot of land in the Amazon the size of a suburban lawn supports 300 species of trees—until it is destroyed.

# HANDS-ON SCIENCE CENTERS WORLDWIDE: SCIENCE MUSEUMS

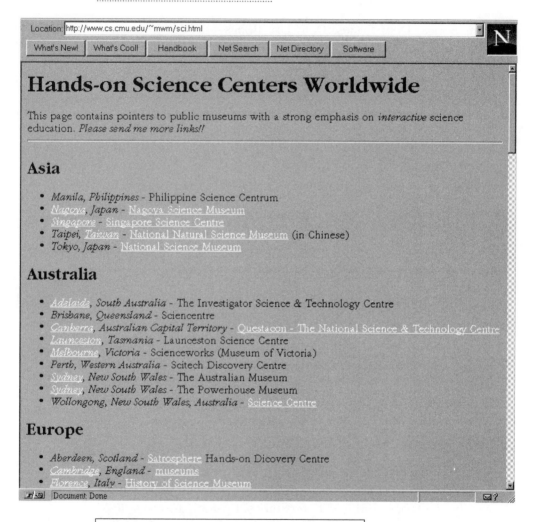

Location: http://www.cs.cmu.edu/~mwm/sci.html

What's New! | What's Cool! | Handbook | Net Search | Net Directory | Software

## Hands-on Science Centers Worldwide

This page contains pointers to public museums with a strong emphasis on *interactive* science education. *Please send me more links!!*

### Asia

- *Manila, Philippines* - Philippine Science Centrum
- *Nagoya, Japan* - Nagoya Science Museum
- *Singapore* - Singapore Science Centre
- *Taipei, Taiwan* - National Natural Science Museum (in Chinese)
- *Tokyo, Japan* - National Science Museum

### Australia

- *Adelaide, South Australia* - The Investigator Science & Technology Centre
- *Brisbane, Queensland* - Sciencentre
- *Canberra, Australian Capital Territory* - Questacon - The National Science & Technology Centre
- *Launceston, Tasmania* - Launceston Science Centre
- *Melbourne, Victoria* - Scienceworks (Museum of Victoria)
- *Perth, Western Australia* - Scitech Discovery Centre
- *Sydney, New South Wales* - The Australian Museum
- *Sydney, New South Wales* - The Powerhouse Museum
- *Wollongong, New South Wales, Australia* - Science Centre

### Europe

- *Aberdeen, Scotland* - Satrosphere Hands-on Dicovery Centre
- *Cambridge, England* - museums
- *Florence, Italy* - History of Science Museum

Document: Done

**http://www.cs.cmu.edu/~mwm/sci.html**

Science Museums have the coolest, most interactive Websites. And here they ALL are, hundreds of them, broken out by continent.

# HISTORY OF SCIENCE, TECHNOLOGY AND MEDICINE

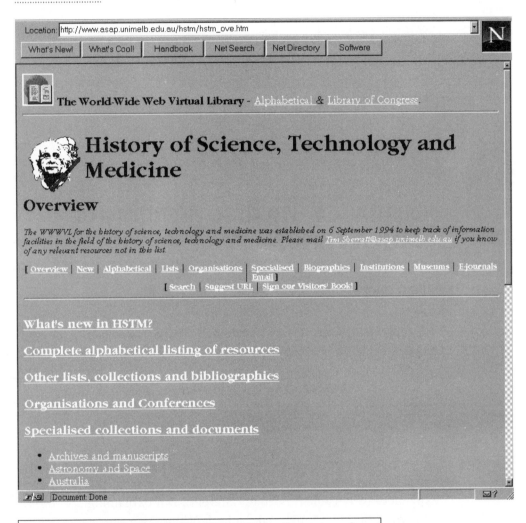

Location: http://www.asap.unimelb.edu.au/hstm/hstm_ove.htm

What's New! | What's Cool! | Handbook | Net Search | Net Directory | Software

The World-Wide Web Virtual Library - Alphabetical & Library of Congress

## History of Science, Technology and Medicine

### Overview

*The WWWVL for the history of science, technology and medicine was established on 6 September 1994 to keep track of information facilities in the field of the history of science, technology and medicine. Please mail Tim.Sherratt@asap.unimelb.edu.au if you know of any relevant resources not in this list.*

[ Overview | New | Alphabetical | Lists | Organisations | Specialised | Biographies | Institutions | Museums | E-journals | Email ]
[ Search | Suggest URL | Sign our Visitors' Book! ]

**What's new in HSTM?**

**Complete alphabetical listing of resources**

**Other lists, collections and bibliographies**

**Organisations and Conferences**

**Specialised collections and documents**

- Archives and manuscripts
- Astronomy and Space
- Australia

Document: Done

**http://www.asap.unimelb.edu.au/hstm/hstm_ove.htm**

Come here for a list of resources related to the history of astronomy, chemistry, computers, scientific instruments, mathematics, medicine, physics, women in the sciences, and more. There are links to museums, journals, document and image collections, bibliographies, and organizations and conferences. You also get links to an on-line biographical dictionary of science and a historical directory of scientific institutions and organizations.

## ION SCIENCE

http://quest.arc.nasa.gov/

Ion Science demystifies complex topics in all branches of science and makes them understandable for children.

## JOURNAL OF STRANGE PHENOMENA: FORTEAN TIMES

http://alpha.mic.dundee.ac.uk/ft/ft.cgi?-1,ft

The *Journal of Strange Phenomena: Fortean Times* is a bimonthly magazine of news, reviews, and research on all manner of strange phenomena and experiences, curiosities, prodigies, and portents. The publication was launched in 1973 to continue the work of the iconoclastic philosopher Charles Fort. Fort was skeptical about scientific explanations. He observed how some scientists argued for and against various theories according to their own beliefs and dogma rather than the rules of evidence. Fort was appalled that data not fitting the collective paradigm were often ignored, suppressed, discredited, or explained away (which is quite different from explaining a thing). If you are of like mind, then take a look at this Web site.

# MICROPATENT

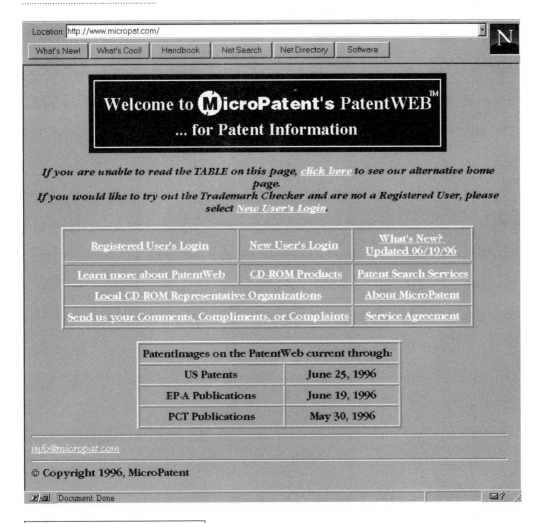

**http://www.micropat.com**

MicroPatent offers Web surfers the ability to search full-text patents by U.S. major classification and/or order the most recent 4 weeks' worth of U.S. patents, search the current week's full-text patents, search last week's full-text patents, or search the front pages of any U.S. patent issues since 1974, or any patent application issued since 1988.

Special downloadable software called PatentImage Viewer allows you to view patent front-page images.

# NATIONAL PUBLIC RADIO: SCIENCE FRIDAY HOT SPOTS

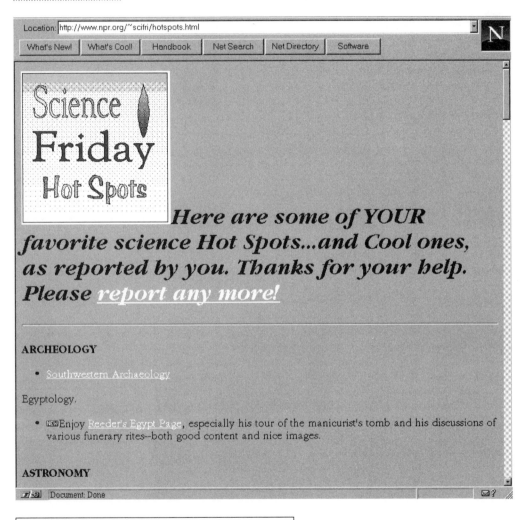

**http://www.npr.org/~scifri/hotspots.html**

This is the superb list of links that was generated during NPR's program on December 29, 1995 discussing science Web sites on the

Internet. The links derive from listener suggestions, and they are arranged alphabetically by subject. You may also access a RealAudio file of the program itself.

# NATIONAL PUBLIC RADIO SCIENCE FRIDAY REALAUDIO ARCHIVE

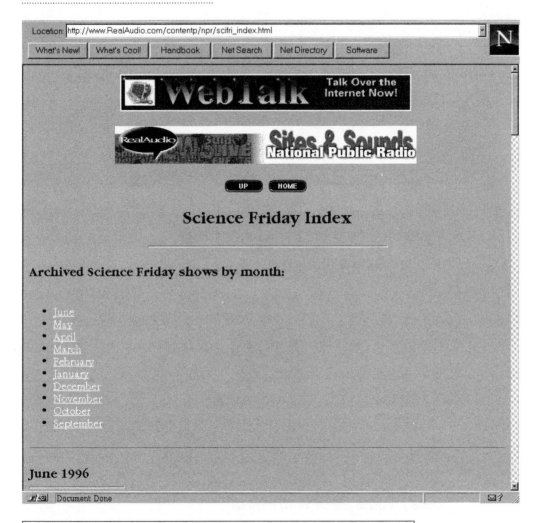

**http://www.RealAudio.com/contentp/npr/scifri_index.html**

Access RealAudio files of NPR Science Friday programs going back more than seven months! Dated abstracts in a table of contents give you an idea of the topics covered on each program.

# NETWORK SCIENCE

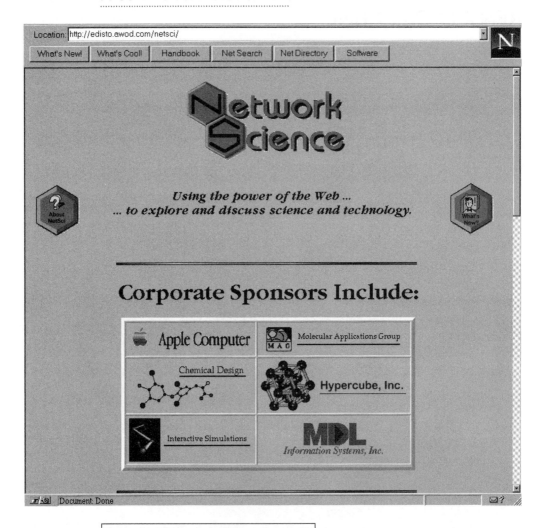

## http://edisto.awod.com/netsci/

Network Science Corporation was founded exclusively to assist in science education by supporting the development of science and technology, by facilitating the timely exchange of information as a means of enhancing professional development, and by fostering public awareness of advances. Topics covered to date include combinatorial chemistry, structure-based drug design, 3-D databases, bioinformatics, and more.

# BILL NYE, THE SCIENCE GUY

http://www.seanet.com/vendors/billnye/nyelabs.html

Visit Bill Nye on-line. Here television's favorite wacky professor provides instructions for experiments, QuickTime video clips, and more as a supplement to his popular television program.

# SciEd: SCIENCE AND MATHEMATICS EDUCATION RESOURCES

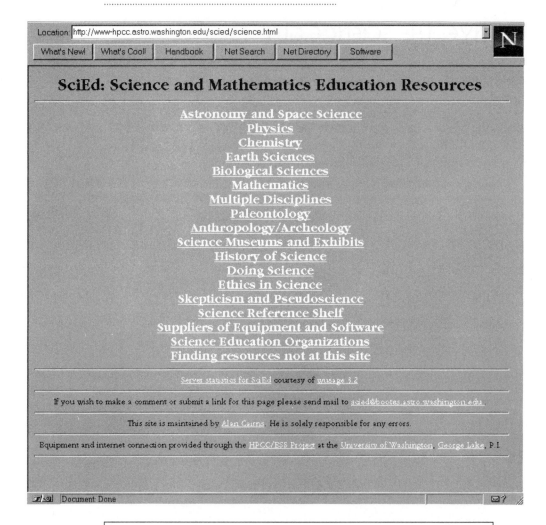

Location: http://www-hpcc.astro.washington.edu/scied/science.html

What's New! | What's Cool! | Handbook | Net Search | Net Directory | Software

## SciEd: Science and Mathematics Education Resources

**Astronomy and Space Science**
**Physics**
**Chemistry**
**Earth Sciences**
**Biological Sciences**
**Mathematics**
**Multiple Disciplines**
**Paleontology**
**Anthropology/Archeology**
**Science Museums and Exhibits**
**History of Science**
**Doing Science**
**Ethics in Science**
**Skepticism and Pseudoscience**
**Science Reference Shelf**
**Suppliers of Equipment and Software**
**Science Education Organizations**
**Finding resources not at this site**

Server statistics for SciEd courtesy of wusage 3.2

If you wish to make a comment or submit a link for this page please send mail to scied@bootes.astro.washington.edu

This site is maintained by Alan Cairns. He is solely responsible for any errors.

Equipment and internet connection provided through the HPCC/ESS Project at the University of Washington, George Lake, P.I.

Document Done

---

**http://www-hpcc.astro.washington.edu/scied/science.html**

Here are hundreds of links arranged by topic and addressing astronomy, physics, chemistry, earth sciences, biological science, mathematics, multiple disciplines, paleontology, science museums and exhibits, history of science, ethics in science, skepticism and pseudoscience, suppliers of equipment and software, and science education organizations.

# SCIENCE HOBBYIST

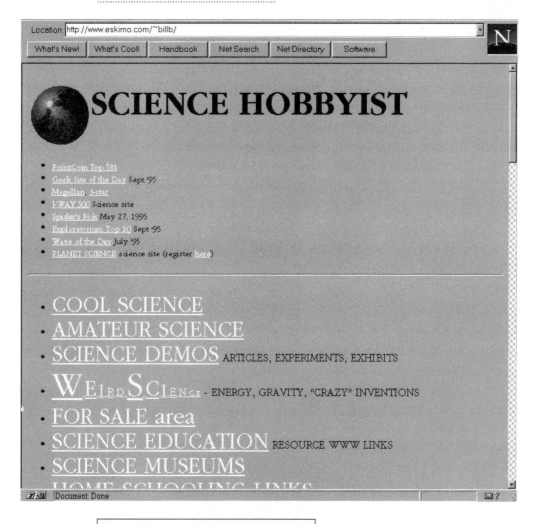

## http://www.eskimo.com/~billb/

The Science Hobbyist Web site offers a vast and wonderful collection of links of use to the amateur scientist, including on-line science demos, experiments, and exhibits.

# SCIENCE ON-LINE

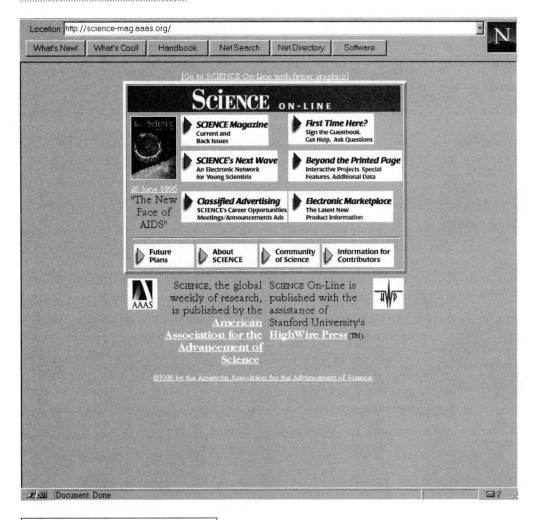

## http://sciene-mag.aaas.org/

Visit the Web home of *Science*, the global weekly of research published by the American Association for the Advancement of Science. Get information on current and back issues. And check out the many interactive projects, special features and additional data that are available beyond the printed page.

# SCIENCE SURF

http://www.well.com/user/wcalvin/scisurf.html

From William H. Calvin, author of *The River That Flows Uphill*, comes this great, eclectic science zine.

# SOCIETY FOR AMATEUR SCIENTISTS

http://www.thesphere.com/SAS/

The Society for Amateur Scientists (SAS) is a nonprofit research and educational organization dedicated to helping people enrich their lives by following their passion to take part in scientific adventures of all kinds. SAS is a unique collaboration between world-class professionals and amateur scientists. SAS is dedicated to helping everyday people find the limits of their own genius by developing their scientific skills and removing the roadblocks that today make it nearly impossible for people without Ph.D. degrees to do research. SAS is a conservative scientific organization dedicated primarily to advancing well-established experimental fields. It does not conduct research into paranormal phenomena. Also, it does not provide support for amateur theories of cosmology, creation, the unified field, or similar topics.

# GEOLOGY

American Geological Institute
American Geophysical Union
Bishop Museum's Geology Page
Canada Geological Survey
Coastal and Marine Geology Projects of the U.S. Geological
    Survey
Earth Science Emporium
Electronic Volcano
Finland Geological Survey
Frequently Asked Questions (FAQs) for sci.geo.geology
Geologic Time Hypertext Reference
Geological Society of America
Induced Earthquake Bibliography

Japan Geological Survey
Margins Initiative Home Page
Michigan Technological University Volcanoes Page
Mineral Gallery
National Geophysical Data Center
Ocean Drilling Program (ODP)
**Pinpoint News**: Earthquake Predictions
Rob's Granite Page
Shell Crustal Imaging Facility
**Southeastern Geology**: A Journal
U.S. Geological Survey
Volcano World
Yale Geology and Geophysics: The Kline Geology Laboratory

# AMERICAN GEOLOGICAL INSTITUTE

http://agi.umd.edu/agi/agi.html

The American Geological Institute (AGI) is a federation of twenty-nine geoscience societies. The AGI's mission is to provide information and education services to its members, promoting a united voice for the geoscience community. Come here to access:

❏ AGI Government Affairs files—updated information on government activities, including important legislation in Congress;

❏ GeoRef—AGI's 1.9-million-record database of essential bibliographic information for the geoscience community;

❏ *GeoTimes* Magazine—featuring news and trends in the geosciences;

❏ AGI employment classifieds for geology professionals.

# AMERICAN GEOPHYSICAL UNION

http://earth.agu.org/kosmos/homepage.html

The American Geophysical Union (AGU) is an international scientific society with more than 32,000 members in over 115 countries. For over seventy-five years, AGU researchers, teachers, and science administrators have dedicated themselves to advancing the understanding of the Earth and its environment. At this site you can access great geoscience articles. A few samples:

❏ "Anticipating the Successor to Mexico's Largest Historical Earthquake,"—which describes the evidence that suggests that a large earthquake should be expected to strike Jalisco, Mexico in the next few years as pressure builds to the breaking point along a major fault in the region.

❏ "Taking Petrologic Pathways Toward Understanding Rabaul's Restless Caldera," which discusses volcanic eruptions that broke out in Rabaul, New Guinea in 1995, for the first time since 1943.

Additionally, you may subscribe to or submit articles for *Earth Interactions*, a great new electronic journal copublished by AGU, the American Meteorological Society, and the Association of American Geographers.

# BISHOP MUSEUM'S GEOLOGY PAGE

http://www.bishop.hawaii.org/bishop/geology/geology.html

Aloha from the Bishop Museum, Hawaii! Because of its volcanic and tectonic setting, Hawaii is an ideal place for geologists to study hot-spot volcanism, mid-ocean ridges, mantle-plume processes, and other geological and volcanological events.

Currently, museum researchers are investigating the structure of the Hawaiian hot spot by studying the lavas erupted at Kilauea over most of its history. They are looking at three very deep cores (1500 to 2000 meters long) that represent 100,000 to 200,000 years of eruptions at Kilauea. This will tell the researchers how the compositions of lavas changed over most of Kilauea's life and from this they will learn about the hot spot's size and structure deep in the Earth's mantle.

Visit the Web site for more details.

# CANADA GEOLOGICAL SURVEY

http://www.emr.ca/gsc/

The mandate of the Geological Survey of Canada is to provide a comprehensive geoscience knowledge base contributing to economic development, public safety, and environmental protection by acquiring, interpreting, and disseminating geoscience information concerning Canada's landmass, including the offshore.

The survey's core responsibilities are to:

❐ map the regional geologic and tectonic framework of Canada's landmass and offshore;

❐ develop an understanding of the nature, quantity, distribution, and formation of Canada's mineral and energy resources;

❐ develop an understanding of the contemporary geological processes affecting Canadian society;

❐ maintain the National Geoscience Database, ensuring that all information is available as maps, reports, etc. in a timely fashion.

And that last responsibility is what this Web site is all about.

# COASTAL AND MARINE GEOLOGY PROJECTS OF THE U.S. GEOLOGICAL SURVEY

http://marine.usgs.gov/fact-sheets/

Here are extensive briefings on and files related to various USGS projects involving Antarctica, Louisiana's Barrier Islands, Louisiana's Lake Pontchartrain Basin and coastal wetlands, geology of the Gulf of the Farallons National Marine Sanctuary, the geology of the Florida keys, geology and the fishery of Georges Bank, and more.

# EARTH SCIENCE EMPORIUM

http://nlu.nl.edu/bthu/nlu/eight/es/Homepage.html

Here are hundreds of links to Web sites related not just to geology, but also to other earth science disciplines including astronomy, meteorology, and oceanography.

# ELECTRONIC VOLCANO

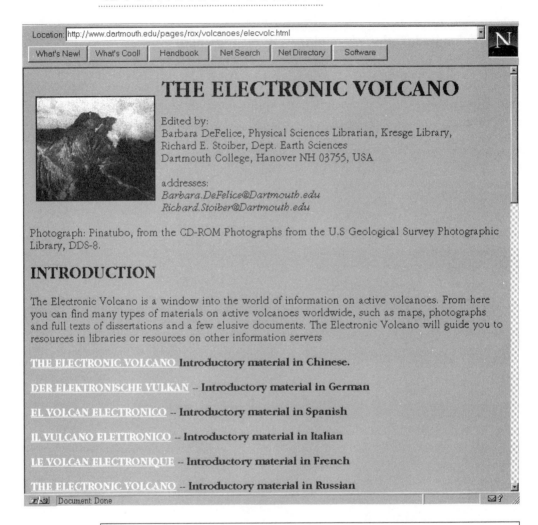

Location: http://www.dartmouth.edu/pages/rox/volcanoes/elecvolc.html

| What's New! | What's Cool! | Handbook | Net Search | Net Directory | Software |

## THE ELECTRONIC VOLCANO

Edited by:
Barbara DeFelice, Physical Sciences Librarian, Kresge Library,
Richard E. Stoiber, Dept. Earth Sciences
Dartmouth College, Hanover NH 03755, USA

addresses:
Barbara.DeFelice@Dartmouth.edu
Richard.Stoiber@Dartmouth.edu

Photograph: Pinatubo, from the CD-ROM Photographs from the U.S Geological Survey Photographic Library, DDS-8.

### INTRODUCTION

The Electronic Volcano is a window into the world of information on active volcanoes. From here you can find many types of materials on active volcanoes worldwide, such as maps, photographs and full texts of dissertations and a few elusive documents. The Electronic Volcano will guide you to resources in libraries or resources on other information servers

THE ELECTRONIC VOLCANO Introductory material in Chinese.

DER ELEKTRONISCHE VULKAN -- Introductory material in German

EL VOLCAN ELECTRONICO -- Introductory material in Spanish

IL VULCANO ELETTRONICO -- Introductory material in Italian

LE VOLCAN ELECTRONIQUE -- Introductory material in French

THE ELECTRONIC VOLCANO -- Introductory material in Russian

Document Done

---

**http://www.dartmouth.edu/pages/rox/volcanoes/elecvolc.html**

The Electronic Volcano is a window into the world of information on active volcanoes. Here you can find many types of materials on active volcanoes worldwide, such as maps, photographs, and full texts of dissertations and other elusive documents. Note that the Electronic Volcano is available not only in English but also in Chinese, Russian,

German, Spanish, Italian, and French. This splendid site is maintained by Barbara DeFelice (Barbara.DeFelice@Dartmouth.edu) and Richard Stoiber (Richard.Stoiber@Dartmouth.edu), both of the Department of Earth Science, Dartmouth College.

# FINLAND GEOLOGICAL SURVEY

http://www.gsf.fi/

The Geological Survey of Finland (GSF) acquires, assesses, and makes available geological information in promoting the balanced, long-term use of natural resources, particularly for the mining and construction industries, as well as for land use, environmental protection, and public health authorities. Founded in 1885, the GSF is funded by the Finnish government.

# FREQUENTLY ASKED QUESTIONS (FAQS) FOR SCI.GEO.GEOLOGY

http://gwrp.cciw.ca/internet/sci.geo.geology/faq.html

Not only do you get the FAQ list from the most popular geology newsgroup on the Net, but also Phillip Ingram's hypertext tutorial entitled *Using the Web for Earth Sciences Information on the Internet.*

# GEOLOGIC TIME HYPERTEXT REFERENCE

http://ucmp1.berkeley.edu/timeform.html

Here is a terrific, information-filled hypertext time line running from the Precambrian to the Cenozoic era (i.e., from the beginning of the

Earth to right this minute). The site also includes a good hypertext introduction to geology.

# GEOLOGICAL SOCIETY OF AMERICA

http://www.aescon.com/geosociety/index.html

Established in 1888, the Geological Society of America has more than 15,000 fellows, members, and student members. Come here for information on meetings, membership, publications, education programs, and more.

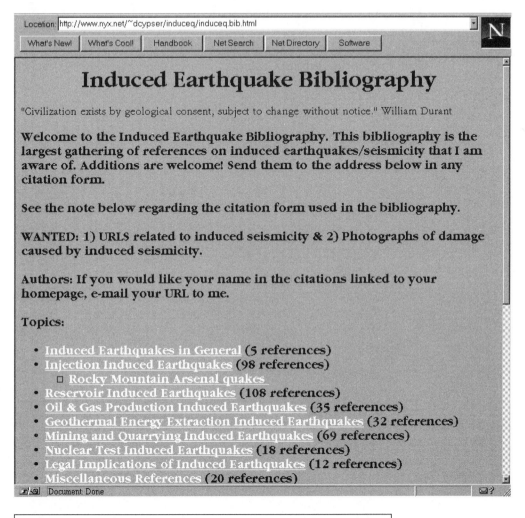

# Induced Earthquake Bibliography

"Civilization exists by geological consent, subject to change without notice." William Durant

Welcome to the Induced Earthquake Bibliography. This bibliography is the largest gathering of references on induced earthquakes/seismicity that I am aware of. Additions are welcome! Send them to the address below in any citation form.

See the note below regarding the citation form used in the bibliography.

WANTED: 1) URLS related to induced seismicity & 2) Photographs of damage caused by induced seismicity.

Authors: If you would like your name in the citations linked to your homepage, e-mail your URL to me.

Topics:

- Induced Earthquakes in General (5 references)
- Injection Induced Earthquakes (98 references)
  - Rocky Mountain Arsenal quakes
- Reservoir Induced Earthquakes (108 references)
- Oil & Gas Production Induced Earthquakes (35 references)
- Geothermal Energy Extraction Induced Earthquakes (32 references)
- Mining and Quarrying Induced Earthquakes (69 references)
- Nuclear Test Induced Earthquakes (18 references)
- Legal Implications of Induced Earthquakes (12 references)
- Miscellaneous References (20 references)

http://www.nyx.net/~dcypser/induceq/induceq.bib.html

Here are links to all you would ever want to know about induced earthquakes, including:

❐ injection-induced earthquakes (88 references)

❐ reservoir-induced earthquakes (91 references)

❏ oil and gas production–induced earthquakes (22 references)

❏ geothermal energy extraction–induced earthquakes (29 references)

❏ mining and quarrying–induced earthquakes (31 references)

❏ nuclear test–induced earthquakes (11 references)

❏ legal implications of induced earthquakes (11 references).

# JAPAN GEOLOGICAL SURVEY

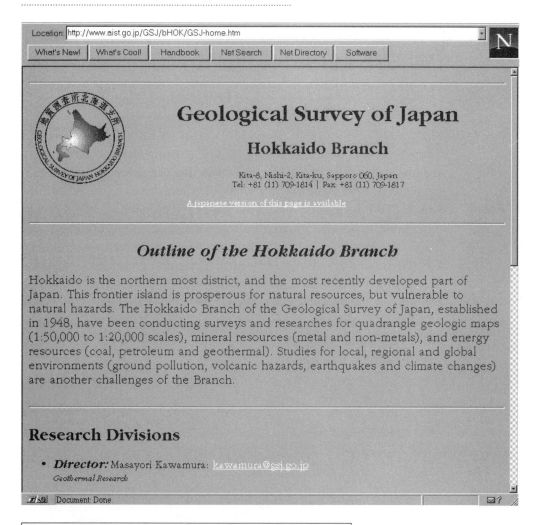

Location: http://www.aist.go.jp/GSJ/bHOK/GSJ-home.htm

| What's New! | What's Cool! | Handbook | Net Search | Net Directory | Software |

## Geological Survey of Japan

### Hokkaido Branch

Kita-8, Nishi-2, Kita-ku, Sapporo 060, Japan
Tel: +81 (11) 709-1814 | Fax: +81 (11) 709-1817

A Japanese version of this page is available

### *Outline of the Hokkaido Branch*

Hokkaido is the northern most district, and the most recently developed part of Japan. This frontier island is prosperous for natural resources, but vulnerable to natural hazards. The Hokkaido Branch of the Geological Survey of Japan, established in 1948, have been conducting surveys and researches for quadrangle geologic maps (1:50,000 to 1:20,000 scales), mineral resources (metal and non-metals), and energy resources (coal, petroleum and geothermal). Studies for local, regional and global environments (ground pollution, volcanic hazards, earthquakes and climate changes) are another challenges of the Branch.

### Research Divisions

- ***Director:*** Masayori Kawamura: kawamura@gsj.go.jp
  *Geothermal Research*

Document Done

**http://www.aist.go.jp/GBJ/bHOK/GSJ-home.htm**

Available in both Japanese and English, this page documents the research of the Hokkaido Branch of the Geological Survey of Japan. Here you will find information on surveys and researches for quadrangle geologic maps, mineral resources (metal and non-metals), and energy resources (coal, petroleum, and geothermal resources). Here you will also find studies of volcanic and earthquake hazards in Japan.

# MARGINS INITIATIVE HOME PAGE

http://zephyr.rice.edu/margins/

*Margins* is a research initiative designed to bring forward concepts of continental margin evolution. The objective is to elevate this science from the present descriptive and largely qualitative characterizations, to a level where theory, modeling, and simulation interact with observation and experiment to yield a self-consistent understanding of the physical, chemical, and biological processes that are the fundamental elements involved in margin formation and evolution. To achieve this objective a group of scientists, with broad community input, has formulated an Initial Science Plan calling for a redefinition of objectives, a redirection of traditional approaches to margin studies, and a reassessment of research strategies. Come here for much more information on the initiative.

# MICHIGAN TECHNOLOGICAL UNIVERSITY VOLCANOES PAGE

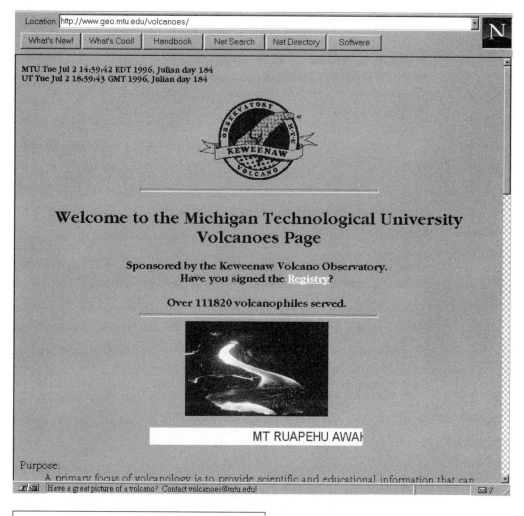

Location: http://www.geo.mtu.edu/volcanoes/

| What's New! | What's Cool! | Handbook | Net Search | Net Directory | Software |

MTU Tue Jul 2 14:59:42 EDT 1996, Julian day 184
UT Tue Jul 2 18:59:43 GMT 1996, Julian day 184

## Welcome to the Michigan Technological University Volcanoes Page

Sponsored by the Keweenaw Volcano Observatory.
Have you signed the Registry?

Over 111820 volcanophiles served.

MT RUAPEHU AWAK

Purpose:
A primary focus of volcanology is to provide scientific and educational information that can

Have a great picture of a volcano? Contact volcanoes@mtu.edu!

---

**http://www.geo.mtu.edu/volcanoes/**

This excellent site includes a clickable worldwide volcanic reference map taking to you to information on specific volcanoes and volcanic regions. You also get information on the most current volcanic activity worldwide as well as details on:

- volcanic hazards mitigation—including historical information, research, images, and discussion of volcanic eruptions and hazards, including data on "dead" volcanoes;

- remote sensing of volcanoes—information about how to process and interpret satellite remote sensing data on volcanoes and eruption clouds;

- terminology and definitions of volcanology—including such arcane terms as *volcaniclastic* and *stratovolcano*;

- and more, including volcanic humor and on-line journals related to volcanology.

# MINERAL GALLERY

http://mineral.galleries.com/default.htm

The Mineral Gallery is a constantly growing collection of mineral descriptions and images, together with several ways of accessing these descriptions. The descriptions include searchable mineralogical data, plus other information of interest to students and other rock hounds. The in-line GIF images (size about 5K) usually are linked to larger JPEG images (averaging 12K). You may search for minerals by name, by class (elements, oxides, carbonates, etc.), by interesting grouping (gemstones, birthstones, etc.), or by keyword.

# NATIONAL GEOPHYSICAL DATA CENTER

http://www.ngdc.noaa.gov/ngdc.html

The National Geophysical Data Center (NGDC) manages environmental data in the fields of marine geology and geophysics, paleoclimatology, solar-terrestrial physics, solid earth geophysics, and glaciology (snow and ice). While not all of NGDC's data holdings are available through their on-line Geophysical Data Center, new data are continually being added.

# OCEAN DRILLING PROGRAM (ODP)

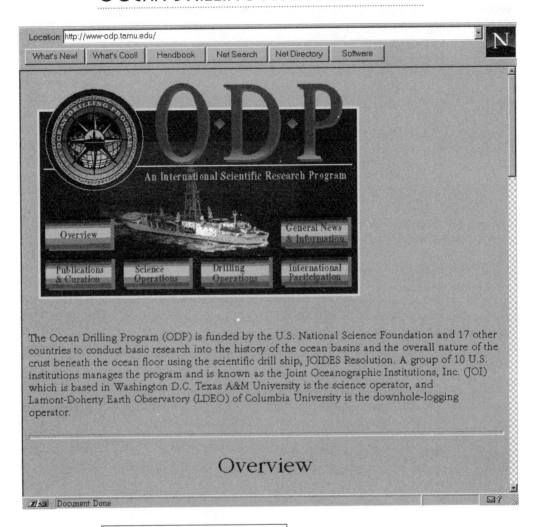

An International Scientific Research Program

Overview

General News & Information

Publications & Curation | Science Operations | Drilling Operations | International Participation

The Ocean Drilling Program (ODP) is funded by the U.S. National Science Foundation and 17 other countries to conduct basic research into the history of the ocean basins and the overall nature of the crust beneath the ocean floor using the scientific drill ship, JOIDES Resolution. A group of 10 U.S. institutions manages the program and is known as the Joint Oceanographic Institutions, Inc. (JOI) which is based in Washington D.C. Texas A&M University is the science operator, and Lamont-Doherty Earth Observatory (LDEO) of Columbia University is the downhole-logging operator.

## Overview

Document: Done

**http://www-odp.tamu.edu**

The Ocean Drilling Program (ODP) is funded by the U.S. National Science Foundation and seventeen other countries to conduct basic research into the history of the ocean basins and the overall nature of the crust beneath the ocean floor using the scientific drill ship, *Joides Resolution*. A group of ten U.S. institutions manages the program and is known as the Joint Oceanographic Institutions, Inc., (JOI), which is based in Washington, D.C. Texas A&M University is the science operator, and Lammont-Doherty Observatory (LDEO) of Columbia University is the downhole-logging operator. Among the resources you'll find here are a schedule of activities for the research ship and an application for research crew participation.

# PINPOINT NEWS: EARTHQUAKE PREDICTIONS

http://www.execpc.com/~bblick/pinpoint

*Pinpont News* is an electronic newsletter currently distributed free to over 300 geologists, seismologists, and other interested parties. The topic is earthquake prediction. Submit your prediction(s), or enter your free e-mail subscription, at this Web site.

# ROB'S GRANITE PAGE

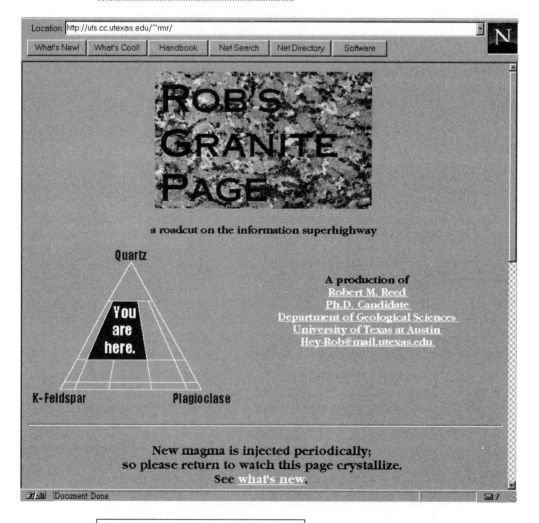

**http://uts.cc.utexas.edu/~rmr/**

This one is hard to describe. The Webmaster calls it "a roadcut on the information highway." New magna is injected periodically; and it is intriguing to watch the page crystallize. Rob is Robert M. Reed (Hey-Rob@mail.utexas.edu), a Ph.D. candidate in the Department of Geological Sciences, University of Texas, Austin.

# SHELL CRUSTAL IMAGING FACILITY

**http://scifwww.scif.uoknor.edu/Home.html**

The Shell Crustal Imaging Facility (SCIF) exists to provide software for seismic reflection processing and interpretation and geophysical research. Check our the resources at this useful Web site.

# SOUTHEASTERN GEOLOGY: A JOURNAL

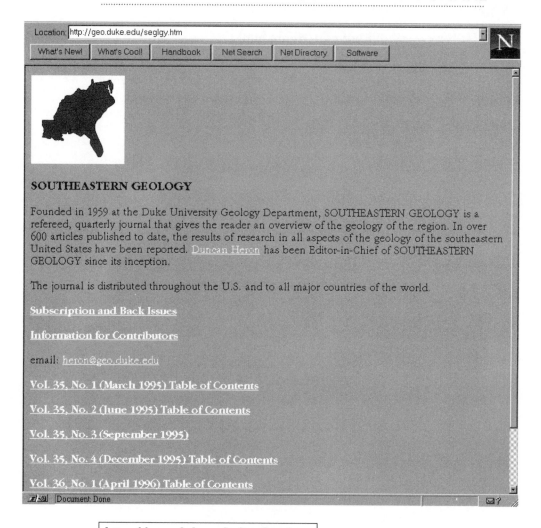

**SOUTHEASTERN GEOLOGY**

Founded in 1959 at the Duke University Geology Department, SOUTHEASTERN GEOLOGY is a refereed, quarterly journal that gives the reader an overview of the geology of the region. In over 600 articles published to date, the results of research in all aspects of the geology of the southeastern United States have been reported. Duncan Heron has been Editor-in-Chief of SOUTHEASTERN GEOLOGY since its inception.

The journal is distributed throughout the U.S. and to all major countries of the world.

Subscription and Back Issues

Information for Contributors

email: heron@geo.duke.edu

Vol. 35, No. 1 (March 1995) Table of Contents

Vol. 35, No. 2 (June 1995) Table of Contents

Vol. 35, No. 3 (September 1995)

Vol. 35, No. 4 (December 1995) Table of Contents

Vol. 36, No. 1 (April 1996) Table of Contents

Document: Done

---

**http://geo.duke.edu/seglgy.htm**

Founded in 1959 at the Duke University Geology Department, *Southeastern Geology* is a refereed, quarterly journal that gives the reader an overview of the geology of the region. In over 600 articles published to date, the results of research in all aspects of the geology of the southeastern United States have been reported. Come here for

TOCs for current and back issues, a fully searchable master index, and subscription information.

# U.S. GEOLOGICAL SURVEY

| http://www.usgs.gov/USGSHome.html |

Access the rich cornucopia of resources provided by the U.S. Geological Survey (USGS). Here you have extensive databases of information on topics in geology, mapping, and water resources. Additionally, there is a complete file of USGS news releases and fact sheets.

# VOLCANO WORLD

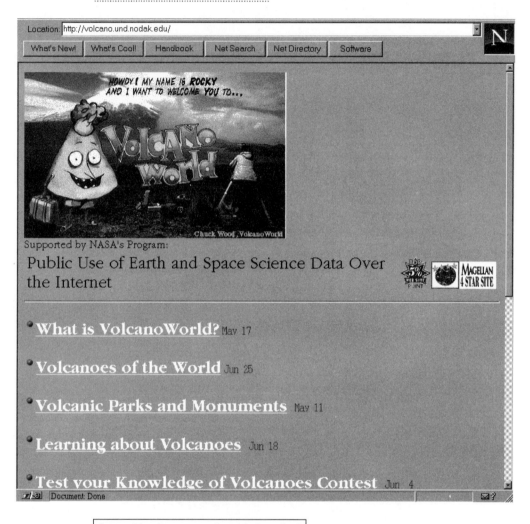

Location: http://volcano.und.nodak.edu/

What's New! | What's Cool! | Handbook | Net Search | Net Directory | Software

Supported by NASA's Program:

## Public Use of Earth and Space Science Data Over the Internet

- **What is VolcanoWorld?** May 17

- **Volcanoes of the World** Jun 25

- **Volcanic Parks and Monuments** May 11

- **Learning about Volcanoes** Jun 18

- **Test your Knowledge of Volcanoes Contest** Jun 4

Document: Done

## http://volcano.und.nodak.edu/

Volcanoes are one of the most dramatic phenomena in nature, attracting millions of visitors each year to U.S. national parks and fascinating millions more schoolchildren in science courses. This page is designed to greatly enrich the learning experiences of these targeted groups by delivering high-quality remote sensing images, other data, and interactive experiments that add depth, variety, and currency to existing volcano information sources.

Volcano World brings modern and near real-time volcano information to specific target audiences and other users of the Internet. Volcano World draws extensively on remote sensing images (AVHRR, Landsat TM, Magellan, Gloria, etc.) and other data collections. The site adds value to these data by relating each image to geologic processes, and by encouraging users to analyze images with provided algorithms.

Volcano World has a very easy to use, Hypercard-like interface. Maps provide an intuitive interface for navigating to further information. Content areas include image-based interactive experiments that stimulate learning because they are fun.

Check out Volcano World.

# YALE GEOLOGY AND GEOPHYSICS: THE KLINE GEOLOGY LABORATORY

http://stormy.geology.yale.edu/kgl.html

Survey the latest Yale research on geochemical archaeometry and archaeometallurgy, geochemistry, geophysical fluid dynamics, mineralogy, petrology and mineral deposits, sedimentation and stratigraphy, solid earth geophysics, and the tectonics of the continental lithosphere.

# MATHEMATICS

# ALGEBRA AND COMPUTING: A BRIEF HISTORY

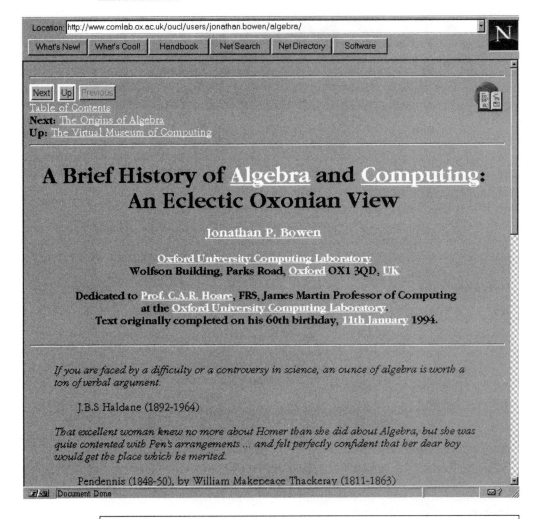

Location: http://www.comlab.ox.ac.uk/oucl/users/jonathan.bowen/algebra/

What's New!   What's Cool!   Handbook   Net Search   Net Directory   Software

Next   Up   Previous

Table of Contents

**Next:** The Origins of Algebra

**Up:** The Virtual Museum of Computing

## A Brief History of Algebra and Computing: An Eclectic Oxonian View

### Jonathan P. Bowen

**Oxford University Computing Laboratory**
Wolfson Building, Parks Road, Oxford OX1 3QD, UK

Dedicated to Prof. C.A.R. Hoare, FRS, James Martin Professor of Computing at the Oxford University Computing Laboratory.
Text originally completed on his 60th birthday, 11th January 1994.

*If you are faced by a difficulty or a controversy in science, an ounce of algebra is worth a ton of verbal argument.*

J.B.S Haldane (1892-1964)

*That excellent woman knew no more about Homer than she did about Algebra, but she was quite contented with Pen's arrangements ... and felt perfectly confident that her dear boy would get the place which he merited.*

Pendennis (1848-50), by William Makepeace Thackeray (1811-1863)

Document: Done

**http://www.comlab.ox.ac.uk/oucl/users/jonathan.bowen/algebra**

This wonderful hypertext tutorial by Jonathan Bowen of the Oxford Computing Laboratory includes discussion of the origins of algebra, early English algebra, algebra and analytical engines, Boolean algebra, and algebra and computing. Bowen has provided a wonderful service to students and colleagues. To thank him, write Jonathan.Bowen@comlab.ox.ac.uk.

# ALGEBRAIC NUMBER THEORY ARCHIVES

http://www.math.uiuc.edu/Algebraic-Number-Theory/

Here is an extensive preprint archive for papers in algebraic number theory and arithmetic geometry. Preprints in electronic form are stored until publication. As is your pleasure, review and read current holdings or submit your own preprints to the archives. The site also includes a remarkably useful search service that allows you to search an index of on-line mathematical texts for key words.

# AMERICAN JOURNAL OF MATHEMATICS

http://muse.jhu.edu/journals/
american_journal_of_mathematics/

Founded in 1878, the *American Journal of Mathematics* is the oldest mathematics journal in the Americas. The journal is published bi-monthly and presents pioneering research papers in the core areas of contemporary mathematics. The full text of several issues are always available on-line, as are guidelines for contributors and back-issue TOCs.

# AMERICAN MATHEMATICAL SOCIETY

http://e-math.ams.org/

The American Mathematical Society was created to further mathematical research and scholarship. Founded in 1888, it now has over

30,000 members, including mathematicians throughout the United States and around the world. At this Web site you will find all the various journals of the society available in hypertext editions, as well as complete membership information.

# CALCULUS AND MATHEMATICA

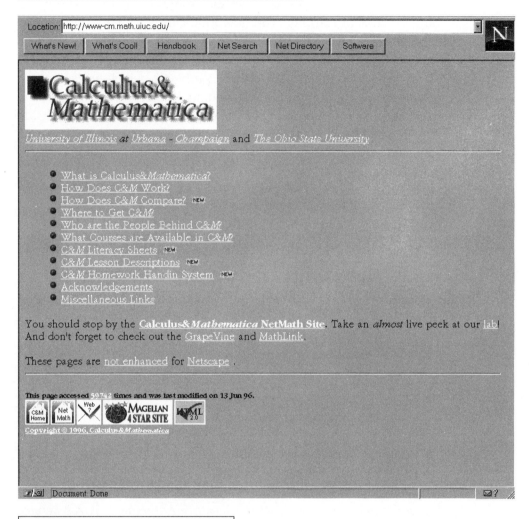

**http://www-cm.math.uiuc.edu/**

Calculus and *Mathematica* is an innovative calculus course that uses the full symbolic, numeric, graphic, and text capabilities of a powerful computer algebra system (*Mathematica*®). Significantly, there is no textbook for this course—only a sequence of electronic notebooks. Each notebook begins with basic problems introducing the new ideas, followed by tutorial problems on techniques and applications. Both problem sets have "electronically active" solutions to support student learning. The notebook closes with a section called "Give It a Try" in which no solutions are given. Students use both the built-in word processor and the graphic calculating software to build their own notebooks to solve these problems, which are submitted electronically for comments and grading. These notebooks have the versatility to allow reworking of examples with different numbers and functions, to provide for the insertion of commentary to explain concepts, to incorporate graphs and plots as desired by students, and to launch routines that extend the complexity of the problem. Come to the Calculus and *Mathematica* Web site for more information. And for more information on *Mathematica*®, see the Wolfram Research Web site at the end of this section.

# CLASSIFICATION SOCIETY OF NORTH AMERICA

**http://www.pitt.edu/~csna/**

The Classification Society of North America is a nonprofit interdisciplinary organization whose purposes are to promote the scientific study of classification and clustering (including systematic methods of creating classifications from data), and to disseminate scientific and educational information related to its fields of interests. Come to this Web site for a searchable index of the society's *Journal of Classification*, for the *Classification Literature Automated Search Service* (*Class*), and for a great, searchable sequence analysis and comparison bibliography prepared by William H.E. Day.

# CLUSTERING SOFTWARE ARCHIVE

http://www.pitt.edu/~csna/Milligan/readme.html

Glenn Milligan provides a splendid archive of downloadable software for cluster generation, hierarchical clustering and inference detection, and k-means clustering with influence detection.

This software includes:

❑ CLUSTGEN.EXE—a DOS-based execute file for cluster generation

❑ CLUSTGEN.FOR—Fortran source code for the cluster generator

❑ HCINFLU.EXE—a DOS-based execute file for hierarchical clustering

❑ HCINFLU.E—Fortran source code for clustering with influence detection

❑ KMINFLU.EXE—a DOS-based execute file for k-means clustering

❑ KMINFLU.F—Fortran source code for clustering with influence detection.

# COMPUTING AND MATHEMATICAL SCIENCES RESEARCH SOFTWARE ARCHIVES AT AT&T

http://netlib.att.com

This is an extensive resource. Here you have downloadable programs of all kinds including:

❑ algorithms for numerical approximation

❑ netlib access tools, such as unshar

- ❏ programs collected from Alliant users worldwide

- ❏ Bessel functions of complex argument

- ❏ toolboxes for linear and nonlinear mathematical programming

- ❏ programs collected from Apollo users

- ❏ benchmark programs

- ❏ biharmonic equations in rectangular geometry and polar coordinates

- ❏ basic linear algebra communication subprograms

- ❏ Brent's multiple precision package

- ❏ miscellaneous codes written in C and C++

- ❏ a parameter program associated with conformal mapping

- ❏ code for generation of examples of continuous-time algebraic Riccati equations

- ❏ a collection of Fortran subroutines that compute the eigenvalues and eigenvectors of nine classes of matrices

- ❏ programs that perform fast Fourier transforms for both complex and real periodic sequences and certain other symmetric sequences

- ❏ and much, much, much more.

# DAVE'S MATH TABLES

http://www.sisweb.com/math/tables.htm

Here are hypertext tables for a host of mathematical topics including:

- general math—number notation

- algebra—basic identities, polynomials, exponents, algebra graphs

- geometry—areas, volumes, surface areas

- trig—identities, tables, hyperbolics, trig graphs

- spherical trig—constants, vectors, and complexity

- calculus—integrals, derivatives

- advanced topics including Fourier series and transforms.

# ELECTRONIC PRIMER ON GEOMETRIC CONSTRAINT SOLVING

http://www.cs.purdue.edu/homes/cmh/electrobook/intro.html

Geometric constraint solving has applications in many different fields, such as molecular modeling, computer-aided design, tolerance analysis, and geometric theorem proving. In this electronic hypertext primer you will find detailed a solution to the problem of finding a configuration for a set of geometric objects that satisfies a given set of constraints between geometric elements. The authors of this tutorial provide four different tours through the material in order to allow different types of users easy access to the information appropriate to their needs. The tutorial includes marvelous two-dimensional constraint solver software.

# FAVORITE MATHEMATICAL CONSTANTS

http://www.mathsoft.com/asolve/constant/constant.html

All numbers are not created equal. That certain constants appear at all and then echo throughout mathematics, in seemingly independent ways, is a source of fascination. Just as physical constants provide "boundary conditions" for the physical universe, mathematical constants somehow characterize the structure of mathematics. The constants itemized at this Web site are rather arbitrarily organized by topic, but are eminently useful. Here you will find all the common, well-known constants as well as constants associated with number theory, analytic inequalities, approximation of functions, enumerating discrete structures, function iteration, and geometry.

# GUIDE TO AVAILABLE MATHEMATICAL SOFTWARE

http://gams.nist.gov/

Here is a great search engine for finding freely downloadable, math-oriented software on the Web. You may search either by the problem addressed by various software or by specific software name.

# MATCOM LIBRARY

http://techunix.technion.ac.il/~yak/matcom.html

MATCOM is a Matlab to C++ compiler and matrix library. The translator creates C++ code from Matlab code, which is then compiled into an executable program. The C++ MATCOM Library supports high-level, Matlab-like syntax, so functions can be conveniently hand-cooled. Matlab algorithms may be included in C++ projects using auto translation. This freeware is available in versions for either Unix or Windows, and both can be downloaded here. The Unix release is a 320-KB compressed .tar file. The Windows release is a 1.3-MB zip file. For the Unix release, note that Unix Netscape may be used to uncompress the file.

# MATHEMATICAL PHYSICS ELECTRONIC JOURNAL

http://www.ma.utexas.edu/mpej/MPEJ.html

The intention of this hypertext journal is to publish papers in mathematical physics and related areas. Research papers and review articles are selected through a normal refereeing process overseen by an editorial board. Visit this Web site to subscribe to the journal, or simply to abstracts. Either subscription is free.

# MATHEMATICAL QUOTATIONS

http://math.furman.edu/~mwoodard.mquot.html

Hundreds of quotations, sorted by source and topic. "Life is good for only two things, discovering mathematics and teaching mathematics," said Simeon Poisson. "The simplest schoolboy is now familiar with facts for which Archimedes would have sacrificed his life," said Ernest Renan. "How dare we speak of the laws of chance? Is not chance the antithesis of all law?" wrote Bertrand Russell.

# MATHEMATICS ARCHIVES

http://archives.math.utk.edu/

The goal of the Mathematics Archives is to provide organized Internet access to a wide variety of mathematical resources. The primary emphasis is on materials that are used in the teaching of mathematics. Currently the Archives are particularly strong in their collection of educational software. Other areas, ranging from laboratory notebooks and problems sets to lecture notes and reports on innovative methods, are growing every day. A second strength of the archives is its extensive collections of links to other sites that are of interest to mathematicians. Resources available through these links include electronic journals, preprint services, grants information, and publishers of mathematical software, texts, and journals.

# MATHEMATICS METASITE

http://euclid.math.fsu.edu/Science/math.html

This Mathematics Metasite provides hundreds of links to general math resources on the Web, math department Web servers worldwide, mathematical software, mathematics gophers and newsgroups, preprints, electronic journals, bibliographies, TeX archives, and more.

# MATHEMATICS ON-LINE BOOKSHELF

http://mathbookshelf.fullerton.edu/

At the Mathematics Online Bookshelf you will find information on mathematics books at all levels. Participating publishers supply the bookshelf with information on their mathematics books. This information is then added to the bookshelf's searchable database, which is updated every week or so. Search by author, title, topic, or publisher. Participating publishers include Academic Press, the American Mathematical Society, Ardsley House, Cambridge University Press, Jones and Bartlett, Kluwer Academic Publishers, Oxford University Press, Plenum Press, Princeton University Press, and the University of Chicago Press.

# MATHEMATRIX, INC.

http://www.primenet.com/~bolster/mmatrix.html

MatheMatrix provides high performance In- and Out-of-Core, Dense, Direct Solve matrix algebra Fortran libraries for both Real Skyline and Complex Full matrices. With MatheMatrix software, Time Stepping and Optimization simulations may also be solved much faster than with LAPACK. Mathematrix libraries are available for various Unix based workstations, MPPs, and Cray YMPs. Come here for more information, including on-line demo sessions.

# MUPAD

http://www-math-uni-paderborn.de/~cube/

MuPAD is a parallel, general purpose computer algebra system developed at the Automath Institut of the University of Paderborn (Germany). Come to this Web page for a demo, and for details on how to acquire a copy of the software via ftp.

# NEW MATHEMATICAL FOUNDATIONS FOR COMPUTER SCIENCE

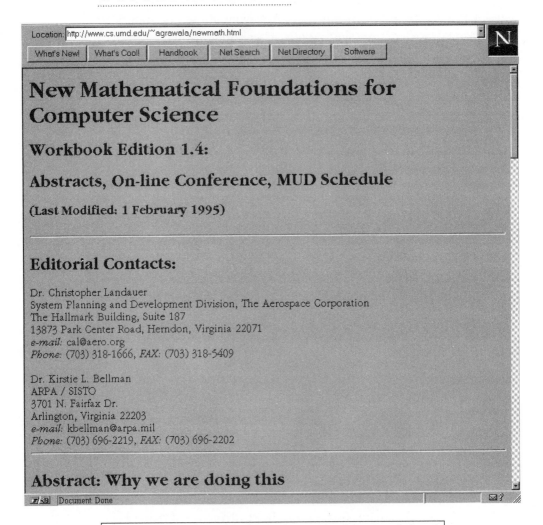

# New Mathematical Foundations for Computer Science

## Workbook Edition 1.4:

## Abstracts, On-line Conference, MUD Schedule

**(Last Modified: 1 February 1995)**

### Editorial Contacts:

Dr. Christopher Landauer
System Planning and Development Division, The Aerospace Corporation
The Hallmark Building, Suite 187
13873 Park Center Road, Herndon, Virginia 22071
*e-mail:* cal@aero.org
*Phone:* (703) 318-1666, *FAX:* (703) 318-5409

Dr. Kirstie L. Bellman
ARPA / SISTO
3701 N. Fairfax Dr.
Arlington, Virginia 22203
*e-mail:* kbellman@arpa.mil
*Phone:* (703) 696-2219, *FAX:* (703) 696-2202

### Abstract: Why we are doing this

---

**http://www.cs.umd.edu/~agrawala/newmath.html**

This Web space is the record of a continuing on-line conference on new mathematical foundations for computer science. The Webmasters intend to collect and redistribute various formal contributions about this topic area. Eventually they expect that all contributors will collectively develop some interesting ideas that will become an edited book of more formal articles. Come here to survey the proceedings, and contribute if you'd like.

# NEWTON'S METHOD

Location: http://www.math.scarolina.edu/cgi-bin/sumcgi/Newton.pl

| What's New! | What's Cool! | Handbook | Net Search | Net Directory | Software |

**Newton's Method**

Equation: `x^2-6*x+2` = 0

x = `1`

**Solve this equation**

Enter an equation in terms of the variable *x*. Then enter a starting guess for *x*. For example, to find the square root of 2 you might enter **x^2-2** for the equation, and **1** as an initial guess. The value after nine iterations is returned in the *x* field and all the iterations can be viewed at the bottom of the page.

Use * for multiplication, / for division, and ^ for exponentiation. You can also use functions such as exp(x), ln(x), cos(x) and sin(x).

Number of Accesses to this page is:

28

Go to the On-line Calculator

Document Done

---

**http://www.math.scarolina.edu/cgi-bin/sumcgi/Newton.pl**

Newton did it by hand. You get an on-line calculator. Enter an equation in terms of the variable *x*. Then enter a starting guess for *x*. For example, to find the square root of 2 you might enter *x^2-2* for the equation and *1* as an initial guess. The value after nine iterations is returned in the $x$ field and all the iterations can be viewed at the bottom of the page.

# NUMBER THEORY WEB

http://www.math.uga.edu/~ntheory/web.html

The Number Theory Web is an attempt to use the resources of the Web to collect and disseminate information of interest to number theorists everywhere and to promote communication between number theorists. Here you will find links to relevant personal home pages and to on-line lecture notes, on-line problem collections, surveys, recent theses, conferences, number theory calculator programs, awards, and more.

# THE PAVILION OF POLYHEDREALITY

http://www.li.net/~george/pavilion.html

Check out George Hart's personal pavilion of practical polyhedra. All the images are copyrighted by George Hart, but you may download and distribute them freely for noncommercial purposes. But that's not all! This is a multimedia event. Hart also provides short musical selections to go with each picture. So if your Web viewer is set up to play MIDI music files, you can listen to Bach as background music while each image is transferring.

Here are the available images, all of them created using POV-Ray:

- ❐ a paper ball made of 3,120 curved arcs (geodesic buffs will notice the lack of alignment along the fivefold axes of symmetry)

- ❐ a puzzle that challenges you to figure out the name of the underlying polyhedron, comprising twelve pentagons and sixty hexagons

- a snub dodecahedron at each vertex of which one pentagon and four equilateral triangles meet

- an artificial radiolarian reticulum.

What are radiolarians, you ask? They are ocean-living protozoa with complex geodesically perforated silica skeletons and sharp spicules. Drawings of a great variety of radiolaria skeletons by the nineteenth century naturalist Ernst Haeckel can be found in many books that deal with geometry, symmetry, and nature. The image here, created by Hart, attempts to capture the idea of a radiolarian but is not faithful to any particular species.

## PERMUTE! 3.2 SOFTWARE

http://alize.ere.umontral.ca/~casgrain/evolution-48-1487.html

Come here to download a free copy of Permute! 3.2, a great Macintosh freeware program to compute regression coefficients and their probability. Here you will also find a self-extracting file packed with data files and documentation for the software.

## PEST SOFTWARE

http://gil.ipswichcity.gld.gov.au/comm/pest/index.html

PEST is software for DOS and UNIX that performs nonlinear estimation for any model. PEST requires no changes or code alternations to a model; the model's source code is not required. The model need not be recompiled because PEST is a parameter estimation shell. It communicates with a model through the model's own input and output files.

PEST can estimate parameters for complex models consisting of multiple executable programs. The number of adjustable parameters and observations are limited only by available memory. Best of all, the program can be used by both programming and nonprogramming modelers to undertake sophisticated calibration or data interpretation tasks.

The software is excellent, but it is not cheap. The DOS version is $500 and the Unix version is $800. But you can get a limited version of the DOS software featuring some, though not all, of the functions as a free download from this site. The limited version is called PESTLITE and it is yours for the taking.

# PROBABILITY WEB

http://www.maths.uq.oz.au/~pkp/probweb/probweb.html

Here are hundreds of links related to all aspects of probability practice and theory.

# R PACKAGE

http://alize.ere.umontreal.ca/~casgrain/R/v3/english

This is freeware for multidimensional analysis and spatial analysis available in Macintosh, VAX, and Windows versions.

# RANDOM NUMBERS AND MONTE CARLO METHODS

http://random.mat.sbg.ac.at/others/

This is a splendid collection of links to random number generators and software for random number generation, as well as cryptography files and software. Here are just a few of your download options:

- ❐ CRAND—a Windows C++ package for transforming realizations of independent, identically distributed variables to various other distributions;

- ❐ ENT—a pseudorandom number sequence test program;

- ❐ RANDPOLY—a REDUCE package based on a port of the Maple random polynomial generator together with some support facilities for the generation of random numbers and anonymous procedures;

- ❐ C++ SIM—an object-oriented Monte Carlo simulation package written in C++.

# REDUCE COMPUTER ALGEBRA SYSTEM

http://www.uni-koeln.de/REDUCE/

REDUCE is an interactive program designed for general algebraic computations of interest to mathematicians, scientists, and engineers. Its capabilities include:

- ❐ expansion and ordering of polynomials and rational functions

- ❐ substitutions and pattern matching in a wide variety of forms

- ❐ automatic and user-controlled simplification of expressions

- [ ] calculations with symbolic matrices

- [ ] arbitrary precision integer and real arithmetic

- [ ] facilities for defining new functions and extending program syntax

- [ ] analytic differentiation and integration

- [ ] factorization of polynomials

- [ ] facilities for the solution of a variety of algebraic equations

- [ ] facilities for the output of expressions in a variety of formats

- [ ] facilities for generating optimized numerical programs from symbolic input

- [ ] calculations with a wide variety of special functions

- [ ] and more.

Come to this Web site for information on the package.

# SEMIGROUPS

http://www.maths.soton.ac.uk/semigroups/homepage.html

Come here for links to home pages of semigroup theorists around the world.

# SOCIETY FOR INDUSTRIAL AND APPLIED MATHEMATICS

http://www.siam.org

The goals of the Society for Industrial and Applied Mathematics (SIAM) are to advance the application of mathematics to science and industry, promote mathematical research that could lead to effective new methods and techniques for science and industry, and provide media for the exchange of information and ideas among mathematicians, engineers, and scientists. SIAM publishes ten peer-reviewed research journals, a new journal reporting on issues and development affecting the applied and computational math community, a quarterly journal of expository and survey papers, and about twenty books per year. Visit the SIAM Web site for more details.

# SOCIETY FOR MATHEMATICAL PSYCHOLOGY

http://www.socsci.uci.edu/smp/

The Society for Mathematical Psychology promotes the advancement and communication of research in mathematical psychology and related disciplines. Mathematical psychology is broadly defined to include work of a theoretical character that uses mathematical methods, formal logic, or computer simulation. Come here for information on the society and its publication, the *Journal of Mathematical Psychology*.

# ST. PATRICK'S COLLEGE, MAYNOOTH, IRELAND, DEPARTMENT OF MATHEMATICS

http://www.maths.may.ie

I have a soft spot for St. Patrick's College in the ancient city of Maynooth, County Kildare, Ireland. A forbear of mine, Laurence F. Renehan (1797–1857), was president here in the mid-nineteenth century and is buried here, where the "old" library is named for him. Today the college boasts one of the finer mathematics departments in Europe. The college celebrated its bicentennial in 1995.

# STATISTICS EDUCATION, AN ELECTRONIC JOURNAL

http://www2.ncsu.edu/ncsu/pams/stat/info/jse/homepage.html

*Statistics Education* is a rigorously refereed electronic journal on post-secondary statistics education. The goal of the journal is to provide interesting and useful information, ideas, software, and data sets to an international readership of statistics educators. The intended audience includes not only members of university statistics departments, but also mathematicians, psychologists, sociologists, and others who teach statistics. All journal materials are available free of charge and can be freely shared among individuals for pedagogical uses.

Topics discussed include curricular reform in statistics, the use of cooperative learning and projects, assessment of students' understanding of probability and statistics and their attitudes and beliefs about statistics, ideas for teaching hypothesis testing, and the use of computing in teaching.

# STATLIB INDEX

http://lib.stat.cmu.edu/

Coming to us courtesy of the Carnegie-Mellon University statistics department, StatLib is a wonderful system for distributing statistical software, data sets, and information by electronic mail, ftp, gopher, and the Web.

Some of the downloadable software you will find here:

❑ CMLIB—This core mathematics library features Fortran subroutines for cluster analysis and related line printer graphics. It includes routines for clustering variables and/or observations using algorithms such as those for direct joining and splitting, Fisher's exact optimization, single-link, k-means, and minimum mutations, as well as routines for estimating missing values.

❑ APSTAT—Selected algorithms, mostly in Fortran, for generating minimal spanning trees, single-link hierarchical clustering, discriminant analysis of categorical data, branch and bound algorithm for feature subset selection, and more.

❑ MULTI—Fortran software for multivariate analysis and clustering, hierarchical clustering, principal components analysis, and discriminant analysis.

There is also general statistical software that includes a 3D interactive data display package, algorithms for convex hull and Delaunay triangulation, and routines for nonlinear discriminant analysis.

# SYNERGETICS ON THE WEB

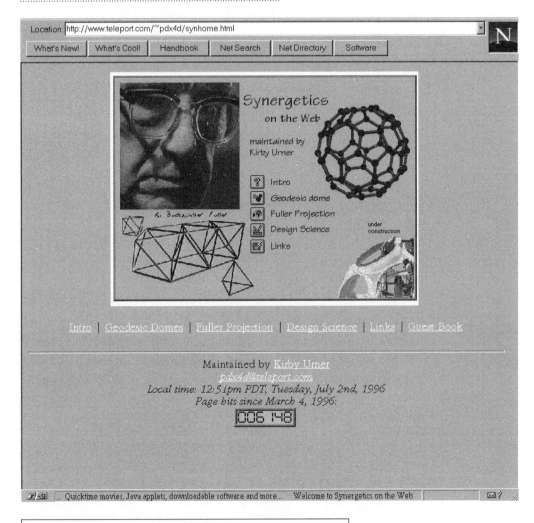

http://www.teleport.com/~pdx4d/synhome.html

Access all manner of information on synergetics, including geodesic domes, Fuller projection, and of course the grand old man himself, Buckminster Fuller.

# TANGENTS ONLINE: THE HARVARD-RADCLIFFE MATHEMATICS BULLETIN

http://www.math.harvard.edu/~hmb/

Come here for an electronic archive of back issues, as well as details on how to subscribe.

# TOOLDIAG: PATTERN RECOGNITION TOOLBOX

http://www.unionova.pt/~tr/home/tooldiag.html

TOOLDIAG is a software toolkit for statistical pattern recognition. The main area of application is classification. The application area is limited to multidimensional continuous features, without any missing values. No symbolic features (attributes) are allowed. The program is implemented in the C language and has been tested in several computing environments. Come to this Web site for more details and a free download.

# VECFEM, VERSION 3

http://www.uni-karlsruhe.de/~vecfem/

VECFEM is a program package for the solution of systems of steady and nonsteady nonlinear partial differential equations (PDE) by the finite element methods. The PDE can be freely defined by the user. The VECFEM code generator creates a Fortran 77 program for the solution of the given PDE. The data structures of VECFEM are optimized for vector computers and parallel computers with distributed memory. Portability is ensured by using Fortran 77. The VECFEM message passing library is automatically mapped into the message passing library of the used parallel computer. The VECFEM kernel runs on PCs, workstations, parallel computers, and vector computers if a Fortran 77 compiler is available. Many more details are to be found at the Web site.

# WAVELETS

http://www.mat.sbg.ac.at/~uhl/wave.html

All you ever wanted to know about wavelets but were afraid to ask. Dozens of great links.

# WOLFRAM RESEARCH

http://www.wri.com

The home page for Mathematica and all of Wolfram's other great products. Check it out. They've even got T-shirts!

# METEOROLOGY

The Daily Planet: Java Weather Visualizer
Digital Atmosphere Software Homepage
Global Change Master Directory

National Oceanic and Atmospheric Administration (U.S.
    Department of Commerce)
Radar and Satellite Web Access
Remote Weather Cameras

# THE DAILY PLANET: JAVA WEATHER VISUALIZER

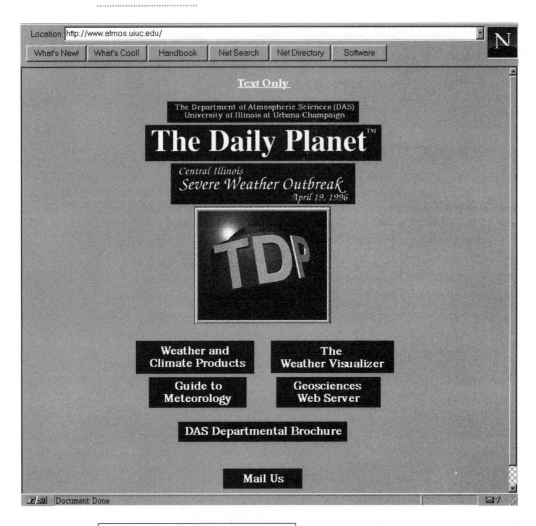

**http://www.atmos.uiuc.edu/**

The Daily Planet is a Web space assembled by the Department of Atmospheric Sciences at the University of Illinois at Champaign-Urbana. The highlight of this site is a weather visualizer map optimized for Hot Java! Among other things, you can customize your own weather maps and images here. By simply pointing and clicking you can select which features you want to see on your customized weather map. Embedded helper files provide useful details and information to provide you with the background necessary to correctly interpret these images. You've got to see it to believe it. And to see it you need a browser configured for Java.

# DIGITAL ATMOSPHERE SOFTWARE HOMEPAGE

**http://ourworld.compuserve.com/homepages/weather/digitala.htm**

What is Digital Atmosphere? It is a software program for Windows that allows you to create, display, and print an infinite variety of weather charts and analyses. It is the same type of tool that is used by National Weather Service offices nationwide. Digital Atmosphere is an absolute must for any serious weather hobbyist, storm enthusiast, pilot, climatologist, or meteorologist. Download a free "lite" version of the software from this Web site. If you like it, then register for a few bucks and get an expanded, more powerful version.

# GLOBAL CHANGE MASTER DIRECTORY

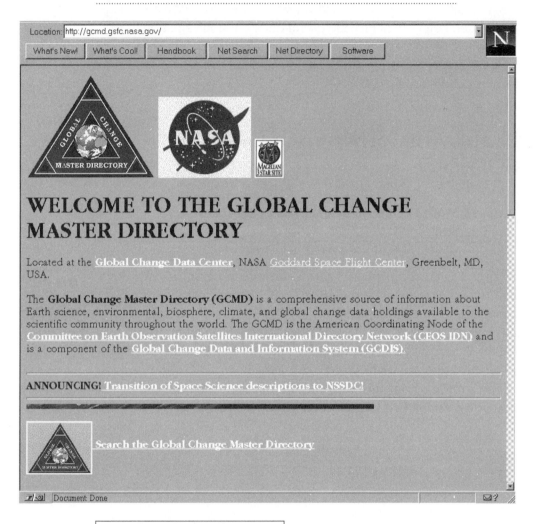

Location: http://gcmd.gsfc.nasa.gov/

| What's New! | What's Cool! | Handbook | Net Search | Net Directory | Software |

## WELCOME TO THE GLOBAL CHANGE MASTER DIRECTORY

Located at the **Global Change Data Center**, NASA Goddard Space Flight Center, Greenbelt, MD, USA.

The **Global Change Master Directory (GCMD)** is a comprehensive source of information about Earth science, environmental, biosphere, climate, and global change data holdings available to the scientific community throughout the world. The GCMD is the American Coordinating Node of the **Committee on Earth Observation Satellites International Directory Network (CEOS IDN)** and is a component of the **Global Change Data and Information System (GCDIS).**

**ANNOUNCING!** Transition of Space Science descriptions to NSSDC!

Search the Global Change Master Directory

Document: Done

## http://gcmd.gsfc.nasa.gov/

Located at the Global Change Data Center, NASA Goddard Space Flight Center, the Global Change Master Directory is a comprehensive source of information about meteorology and related topics. The site features links to hundreds of related resources around the world, including sites linked to earth observation satellites worldwide.

# NATIONAL OCEANIC AND ATMOSPHERIC ADMINISTRATION (U.S. DEPARTMENT OF COMMERCE)

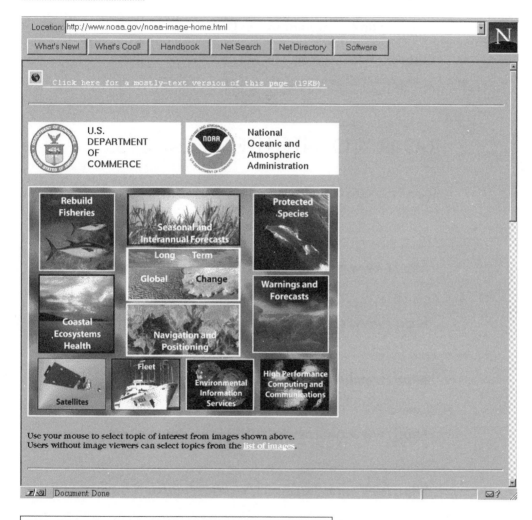

Location: http://www.noaa.gov/noaa-image-home.html

What's New! | What's Cool! | Handbook | Net Search | Net Directory | Software

Click here for a mostly-text version of this page (19KB).

U.S. DEPARTMENT OF COMMERCE

National Oceanic and Atmospheric Administration

Rebuild Fisheries

Seasonal and Interannual Forecasts

Long — Term

Global — Change

Protected Species

Warnings and Forecasts

Coastal Ecosystems Health

Navigation and Positioning

Satellites

Fleet

Environmental Information Services

High Performance Computing and Communications

Use your mouse to select topic of interest from images shown above.
Users without image viewers can select topics from the list of images.

Document Done

**http://www.noaa.gov/noaa-image-home.html**

This site provides access to all the servers of the U.S. National Weather Service, including:

- ❏ the National Hurricane Center

- ❏ the Arkansas-Red Basin River Forecast Center

- ❏ the Climate Prediction Center

- ❏ the Hydrologica Information Center

- ❏ the Midwest Agricultural Weather Service Center

- ❏ and the Spaceflight Meteorology Group.

# RADAR AND SATELLITE WEB ACCESS

Access the Internet's finest Nexrad, regional radar, and color satellite graphics for regions around the globe.

Africa

**http://www.cnn.com/WEATHER/accu.data/afsat.gif**

Antarctica

**http://www.ssec.wisc.edu/data/comp/latest_ant.gif**

Asia

**http://lumahai.soest.hawaii.edu/gifs/gms_cur.gif**

Australia

**http://www.cnn.com/WEATHER/accu.data/aussat.gif**

Canada

**http://www.intellicast.com/weather/intl/cansat.gif**

Caribbean

**http://www.intellicast.com/weather/intl/cbsat.gif**

Europe

**http://www.meteo.fr/tpsreel/images/satt0.gif**

Hawaii

**http://lumahai.soest.hawaii.edu/gifs/hawaii_ir.gif**

Mideast

**http://www.cnn.com/WEATHER/accu.data/Midesat.gif**

New Zealand

**http://www.gphs.vuw.ac.n2/meteorology/pictures/ir1/latest.jpeg**

South America

**http://www.cnn.com/WEATHER/accu.data/samersat.gif**

# REMOTE WEATHER CAMERAS

Wondering what the weather looks like around the world right now? Select a link below to get a semi-live picture of your favorite city or resort. Use these links to travel around the world from your desktop.

Alabama (Huntsville)

**http://www.whnt19.com/towercam/extra.jpg**

Arizona (Phoenix)

**http://www.azfms.com/Travel/camera.html**

Bermuda (St. George's)

**http://www.bbsr.edu/~norm/whatsup.gif**

British Columbia (Vancouver)

**http://wwwmultiactive.com/MFun/cam1.cgi**

http://www.telemark.net/cgi-shl/wc.pl

California (Hollywood)

**http://rfx.rfs.com/images/holly.jpg**

California (Lake Tahoe)

**http://www.rsn.com/~tahoe/**

California (Los Angeles)

**http://www.fountainhead.com/sniper.gif**

California (Mammoth Mountain)

**http://www.rsn.com/~mammoth**

California (San Diego)

**http://www.live.net/sandiego**

California (San Francisco)

**http://www.kpix.com/live/**

http://www.evo.net/bridge

California (Santa Cruz)

**http://sapphire.cse.ucsc.edu/SlugVideo**

Colorado (Boulder)

**http://www.gwha.com/dynimg/snap.jpeg**

Colorado (Colorado Springs)

**http://www.ceram.com/cheyenne/chey.html**

Colorado (Copper Mountain)

**http://www.rsn.com/~copper/**

Colorado (Crested Butte)

**http://www.rsn.com/~crested**

Colorado (Pike's Peak)

**http://www.softronics.com/peak_cam/cam.jpg**

Colorado (Vail)

**http://www.rsn.com/~vail**

Florida (Cape Canaveral)

**http://www.fl-fishing.cam/menu/river.shtml**

Florida (Tampa)

**http://www.wtvt.com/skycam.htm**

Georgia (Atlanta)

**http://vista.homecom.com/cgi-bin/cam2**

Georgia (Savannah)

**http://users.aol.com/wtocwxex/sky.htm**

Hawaii (Manoa Valley)

**http://satftp.soest.hawaii.edu/manoa.html**

Hawaii (Maui)

**http://www.maui.net/~sync/cam.html**

Hawaii (Oahu)

**http://planet-hawaii.com/ph/he.html**

Illinois (Chicago)

**http://www.habitat.com/cgi-bin/show_view**

http://www.wgntv.com/thrlcam.html

Maine (Sugarloaf Mountain)

**http://www.rsn.com/~sloaf/**

Massachusetts (Boston)

**http://www-
1.openmarket.com/boscam/boscam/boscambig.gif**

Michigan (Boyne Mountain)

**http://www.boyne.com/mountain/camera1.html**

Montana (Bozeman)

**http://www.gomontana.com/skycam.jpg**

Nebraska (Lincoln)

**http://www.starcitymall.com/webcam/**

New Hampshire (Mount Washington)

**http://www.rsn.com/~tucks/**

New Hampshire (Wildcat Ski Area)

**http://www.rsn.com/~wildcat/**

New York (Brooklyn Bridge)

**http://romdog.com/bridge/brooklyn.html**

New York (Empire State Building)

**http://205.230.66.5/view.html**

New York (Manhattan)

**http://205.232.5.4/neighbor.nclk**

New York (Pratt University Campus)

**http://www.pratt.edu/prattcam.html**

New York (Rockefeller Center)

**http://www.ftna.com/cam.gif**

Nova Scotia (Halifax)

**http://www.phys.ocean.dal.ca/video/camera.html**

Ohio (Toledo)

**http://www.mco.edu/comp/camalli.shtml**

Oklahoma (Tulsa)

**http://kjrh.com/images/default3.gif**

Ontario (Niagara Falls)

**http://FallCam.niagara.com/FallsCam/Live/Movies/Falls.jpg**

Oregon (Portland)

**http://www.tek.com/PDX_Pictures/pictures.gif**

Pennsylvania (Pittsburgh)

**http://goober.graphics.cs.cmu.edu/~ajw/goober.gif**

Tennessee (Memphis)

**http://www.wmcstations.com/weather/high5m.shtml**

Texas (College Station)

**http://ENTCWEB.tamu.edu/camera.htm**

Utah (Salt Lake City)

**http://www.net.utah.edu/html/cameras.html**

Washington, DC

**http://wxnet4.nbc4.com/cgi-bin/showScrCap**

Washington State (Seattle)

**http://www.cac.washington.edu:1180/cambots/**

West Virginia (Morgantown)

**http://www.dmssoft.com/live.htm**

# OCEANOGRAPHY

Acoustical Oceanography Research Group
Australian Oceanographic Data Center
Center for Coastal Physical Oceanography: Old Dominion
    University
Center for Marine Science Research, University of North
    Carolina, Wilmington
Coral Reefs and Mangroves: Modeling and Management
Monterey Bay Aquarium Research Institute
National Marine Fisheries Service
National Museum of Natural History: Ocean Planet Exhibit
National Weather Service Buoy and CMAN Information
WeatherNet

Ocean Research Institute, Tokyo
Oceanic Information Center, University of Delaware
Pacific Marine Environmental Laboratory
Polar Science Center
Regional Association for Research on the Gulf of Maine
Rosenthal School of Marine and Atmospheric Science
Rutgers Institute of Marine and Coastal Sciences
Scripps Institution of Oceanography
South African Data Center for Oceanography
Space Oceanography Group at Johns Hopkins
Victoria University Center for Earth and Ocean Research
Woods Hole Oceanographic Institution

# ACOUSTICAL OCEANOGRAPHY RESEARCH GROUP

http://pinger.ios.bc.ca/

The oceans are nearly transparent to sound and nearly opaque to light and radio waves. For example, at a wavelength of 1 meter water is nearly one million times more transparent to sound than to radio signals. This fact underlies the intense interest now being directed toward the acoustical exploration of the ocean. The sound waves now being investigated for their potential in ocean science range from wavelengths of a millimeter to over a kilometer.

This new science is concerned both with the interpretation of natural sounds of the sea and with use of active acoustical sources. Natural sounds include the sound of breaking waves and the impact of flying spray, of rainfall, cracking sea ice, the movement of gravel on the sea floor, the sound of submarine earthquakes and volcanoes, and the voices of marine mammals and other sea life.

Our ability to use these signals that nature provides depends on our understanding of the way in which they are created and how they are changed as they travel through the ocean. Mathematical models help us to interpret the naturally occurring sound and to use it to learn more about the processes that give rise to it.

Active sound sources are used to prove the structure of the ocean in two different ways. A sound can be transmitted at one place and received at another. Sounds can be heard thousands of kilometers away. As they travel through the ocean they are modified in one way or another. Thus scientists can use the signals to learn more about the structure and properties of the ocean.

Sound can also be reflected back to its source. The sea floor, the ocean surface, bubbles, fish, plankton and temperature structure can all contribute to scattering. Thus scientists can use sonar systems like search lights to probe the ocean and reveal its properties.

The acoustical Oceanography Research Group is part of the Institute of Ocean Sciences, a research facility of the Canadian Department of Fisheries and Oceans, located on Canada's west coast. The group is dedicated to the development and use of acoustic methods to help understand all aspects of the ocean. This involves both the engineering development of new instruments, and the scientific use of these instruments to answer oceanographic questions.

# AUSTRALIAN OCEANOGRAPHIC DATA CENTER

http://www.aodc.gov.au/AODC.html

The Australian Oceanographic Data Center acquires, manages, and disseminates marine environmental data to the civilian marine science community, the general public, and the Australian Defense Forces.

# CENTER FOR COASTAL PHYSICAL OCEANOGRAPHY: OLD DOMINION UNIVERSITY

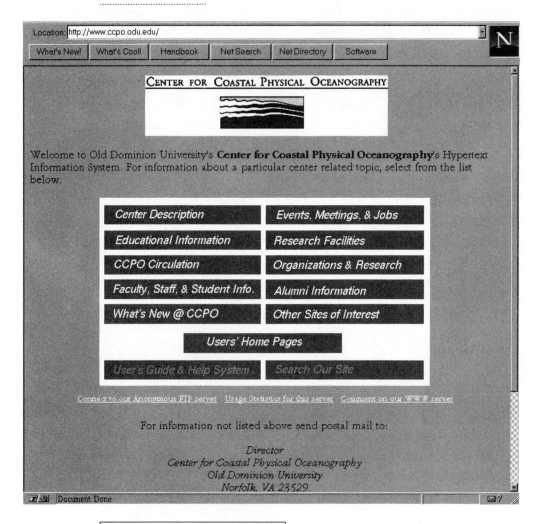

http://www.ccpo.odu.edu/

The Center for Coastal and Physical Oceanography (CCPO) at Old Dominion University was established July 1, 1991. Designed as a center of excellence by the State Council of Higher Education for Virginia, CCPO reflects the Commonwealth of Virginia's commitment to exploring the effects of short-term human impact and long-term global change on the coastal ocean environment. The goal of CCPO is to facilitate research on the physics of the coastal ocean and to provide educational experiences for the public.

Research conducted at CCPO includes both observational and modeling activities. A major task is to develop comprehensive and predictive models for the Atlantic coastal oceans and adjacent shelf regions. These models incorporate important processes influencing shelf circulation, such as wind, river inflow, and offshore currents, and are capable of predicting the behavior of coastal water in regions as vast as the east coast of North America. Important practical applications of this research include predicting crab larvae dispersal and determining how the ocean affects the trajectories of oil spills.

The center performs cooperative research agreements with appropriate federal, state, and private research organizations to promote joint research in areas of mutual interest. As part of its outreach program, CCPO hosts annual workshops arranged on topics of interest to federal and state scientists and environmental managers. Visit the Web site for more details.

# CENTER FOR MARINE SCIENCE RESEARCH, UNIVERSITY OF NORTH CAROLINA, WILMINGTON

http://www.uncwil.edu/sys$disk1/cmsr/cmsr.html

The Center for Marine Science Research is dedicated to providing environment that fosters a multidisciplinary approach to questions in basic marine research. The mission of the center is to promote basic and applied research in the fields of oceanography, coastal and wetlands studies, marine biomedical and environmental physiology, and marine biotechnology and aquaculture. Access this Web site to review the work of faculty members conducting marine science research in the departments of biological sciences, chemistry, and earth sciences. This work includes:

❑ oceanographic research—focusing on continental shelf and slope processes, the integration of estuaries and the near shore continental shelf, and the effects of coastal development on water quality, recruitment, nursery ground degradation, and groundwater;

❑ marine biomedical and environmental physiology research—focusing on developing the use of marine organisms as sources of new products and information;

❑ coastal and estuarine systems research—focusing on the study of the complex interrelationships between barrier beaches, adjacent estuaries, and ocean waters

❑ and more.

# CORAL REEFS AND MANGROVES: MODELING AND MANAGEMENT

http://ibm590.aims.gov.au/

Mangroves and coral reefs are being destroyed or degraded at an alarming rate. The problem is exacerbated by the lack of communication between oceanographers, biologists, and resource managers. To solve this problem, in early 1994 an oceanography team from the Australian Institute of Marine Science (AIMS) merged forces with IBM to improve management of Australia's Great Barrier Reef through electronic communication. This Web site, which includes great 3-D models of water circulation around the Great Barrier Reef, is part of the result. Check it out.

# MONTEREY BAY AQUARIUM RESEARCH INSTITUTE

http://www.mbari.org/

A unique private oceanographic center, the nonprofit Monterey Bay Aquarium Research Institute was established by David Packard with the goal of developing state-of-the-art equipment, instrumentation, systems, and methods of scientific research in the deep waters of the ocean.

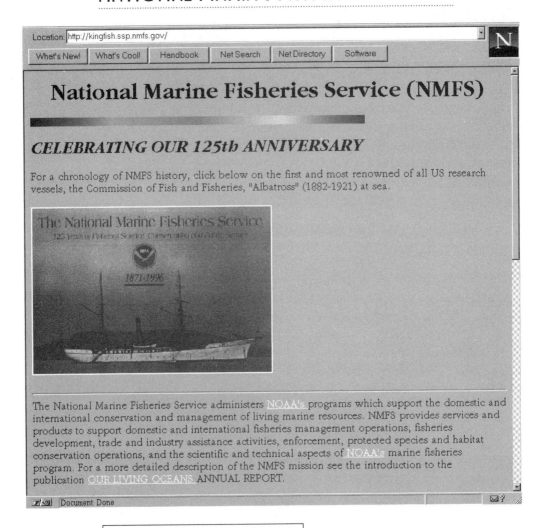

**http://kingfish.ssp.nmfs.gov/**

The National Marine Fisheries Service (NMFS) administers those programs of the National Oceanic and Atmospheric Administration (NOAA) that support the domestic and international conservation and management of living marine resources. NMFS provides products and services to support domestic and international fisheries management operations, fisheries development, trade and industry assistance activities, enforcement, protected species and habitat conservation operations, and the scientific and technical aspects of NOAA's marine fisheries program.

# NATIONAL MUSEUM OF NATURAL HISTORY: OCEAN PLANET EXHIBIT

http://seawifs.gsfc.nasa.gov/ocean_planet.html

The culmination of a four-year effort to study and understand environmental issues affecting the health of the world's oceans, Ocean Planet employs cutting-edge technology, compelling objects and photos, enticing text and walk-through environments to promote celebration, understanding, and conservation of the world's ocean. The exhibit is great and the Web site is almost as good. Check it out.

# NATIONAL WEATHER SERVICE BUOY AND CMAN INFORMATION

http://www.met.fsu.edu/~nws/buoy

Get real-time meteorological and oceanographic data from buoys and CMAN stations all around the United States, the Caribbean, the Gulf of Mexico, both the east and west coasts, and Hawaii and the western Pacific.

# WEATHERNET

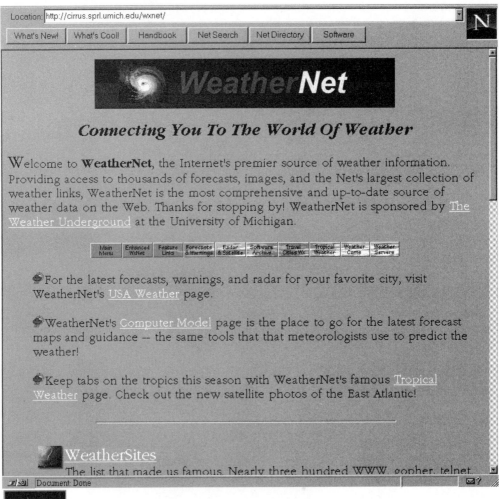

## Connecting You To The World Of Weather

Welcome to **WeatherNet**, the Internet's premier source of weather information. Providing access to thousands of forecasts, images, and the Net's largest collection of weather links, WeatherNet is the most comprehensive and up-to-date source of weather data on the Web. Thanks for stopping by! WeatherNet is sponsored by The Weather Underground at the University of Michigan.

| Main Menu | Enhanced WxNet | Feature Links | Forecasts & Warnings | Radar & Satellite | Software Archive | Travel Cities Wx | Tropical Weather | Weather Cams | Weather Servers |

For the latest forecasts, warnings, and radar for your favorite city, visit WeatherNet's USA Weather page.

WeatherNet's Computer Model page is the place to go for the latest forecast maps and guidance -- the same tools that that meteorologists use to predict the weather!

Keep tabs on the tropics this season with WeatherNet's famous Tropical Weather page. Check out the new satellite photos of the East Atlantic!

WeatherSites
The list that made us famous. Nearly three hundred WWW, gopher, telnet,

Document: Done

**http://cirrus.sprl.umich.edu/wxnet/**

WeatherNet offers one-stop shopping for all kinds of weather and weather-science information. The site includes not only hyperlinks to related Web sites, but also connections to newsgroups, mailing lists, and gophers.

Most importantly, the site provides lots of great PC and Mac shareware available for free downloading, such as:

❏ HurrTrk v. 6.0 for Windows—HurrTrk has long been a favorite among tropical weather buffs. The software features dozens of innovative tools and features, such as zoomable tracking maps. HrrTrk makes it easy to chart hurricanes as they meander across the Atlantic. Historic data for over a dozen famous storms is included with the package, as well as many other goodies.

❏ WeatherGraphix v 4.1 for DOS—Turn your lifeless surface and upper air data into colorful, high-resolution maps. Surface charts, constant pressure analyses, and radar composites are just some of the exciting and useful products you can generate with this powerful software. Note that a 386 hard drive or better, and either a VGA or EGA graphics driver are needed.

❏ Merlin v2.0a for Windows—Track and forecast the movement of hurricanes with this powerful new application by Ted Parker. What distinguishes Merlin from other hurricane tracking software is its utilization of artificial intelligence and an advanced mathematical model. Despite its power, Merlin is extremely easy to use.

❏ RAOB v 3.1 for DOS—Analyze upper-air conditions with ease using RAOB, a new program by John Shewchuk. RAOB instantly converts raw alphanumeric data into high-resolution maps and charts. Icing, turbulence, severe weather data, hail, and lightning are among the parameters you can investigate. This shareware, which requires VGA, is great for pilots and severe weather buffs.

❏ Go Canes! for DOS—This new hurricane tracking software features customizable high-definition maps, storm tagging, animated plots, and command line capability for fast plots. Histories of famous storms are included in the package. Color VGA is required and fast disk/cache is highly recommended.

❏ WxView v 2.8 for DOS—This one is a favorite among weather enthusiasts. WxView converts your raw National Weather Service data into exciting maps and tables. The software is packed with features that make analyzing weather data fast and enjoyable, even for novices.

❏ Sharp for DOS—Similar to RAOB, Sharp converts raw alphanumeric data into upper air analyses.

❏ Quik-Sky v 4.6 for DOS—A fun and easy-to-use weather mapping application that generates graphics from raw alphanumeric data.

❏ CALMET Metar Tutorial for DOS—Shortly, the U.S. National Weather Service will adopt a new surface observation coding scheme called Metar. CALMET's Metar tutorial software prepares you for the change through a series of well-designed lessons and dialogue boxes.

❏ WinWeather v 1.0 for Windows—WinWeather brings hourly weather reports and forecasts right to your PC. Using your SLIP/PPP connection, you'll be able to get the latest weather reports for cities all over the United States and around the world, with no need for you to remember long Internet addresses.

❏ Tracking the Eye v 1.7 for Windows—Here is excellent software for tracking severe storms. Tracking the Eye includes features like toolbars, status bars, graphs, full-color printing, print previewing, and context-sensitive help. The software can instantly determine the distance between any city on the eastern seaboard and the storm it is tracking, and can import ASCII data from other storm trackers. It also features animation, sounds, and statistical graphs. What more can you ask?

❏ Weather Canada v 2.0 for Windows—This is a simple Winsock-compliant program that automatically downloads the latest forecasts for a number of cities across Canada.

❏ Blue-Skies v 1.1 for Mac and Blue-Skies v 1.1 for PowerMac—Bring your weather maps to life with Blue-Skies v 1.1, the innovative new software from the University of Michigan Weather Underground! Features include interactive satellite maps and a multimedia interface that make finding weather data on the Net easy and fun. This is perfect for K-12 earth science programs. The software requires a SLIP/PPP connection or direct Internet connection.

❐ MacWeather v 2.04 for Mac—Retrieving data from the University of Michigan Weather Underground has never been easier! With MacWeather v 2.04, the latest forecasts for and conditions in over 1,000 cities nationwide are just a mouse click away. MacWeather automatically tracks down the data you request and displays them in graphical form. The software requires a SLIP/PPP account or a direct Internet connection.

# OCEAN RESEARCH INSTITUTE, TOKYO

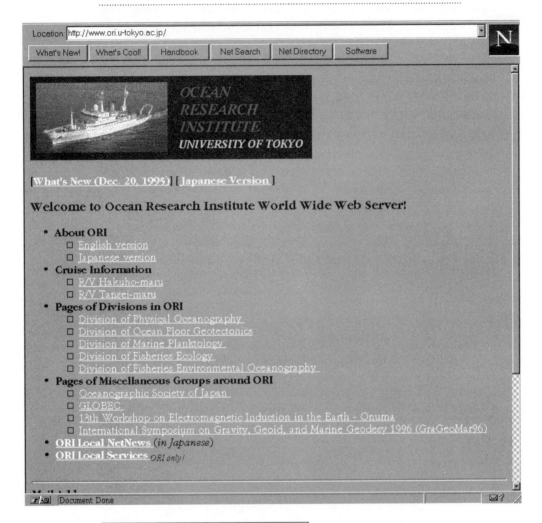

**http://www.ori.u-tokyo.ac.jp/**

The Ocean Research Institute (ORI) of the University of Tokyo was established in April 1962. ORI is comprised of sixteen research departments on the Nakano campus, the Otsuchi Marine Research Center, and the new established Center for International Cooperation. Get details on research and programs by visiting the Web site.

# OCEANIC INFORMATION CENTER, UNIVERSITY OF DELAWARE

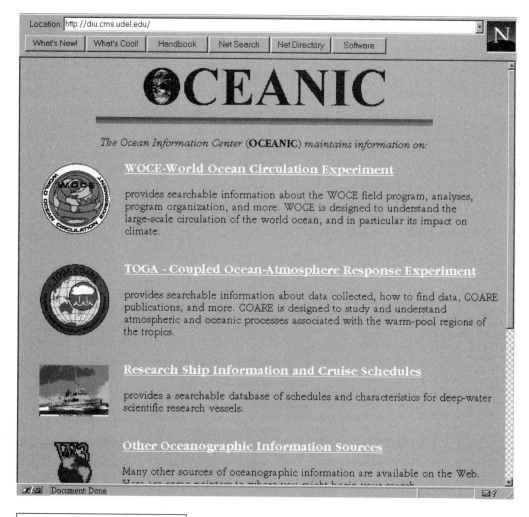

Location: http://diu.cms.udel.edu/

What's New! | What's Cool! | Handbook | Net Search | Net Directory | Software

# ⊕CEANIC

The Ocean Information Center (**OCEANIC**) maintains information on:

## WOCE-World Ocean Circulation Experiment

provides searchable information about the WOCE field program, analyses, program organization, and more. WOCE is designed to understand the large-scale circulation of the world ocean, and in particular its impact on climate.

## TOGA - Coupled Ocean-Atmosphere Response Experiment

provides searchable information about data collected, how to find data, COARE publications, and more. COARE is designed to study and understand atmospheric and oceanic processes associated with the warm-pool regions of the tropics.

## Research Ship Information and Cruise Schedules

provides a searchable database of schedules and characteristics for deep-water scientific research vessels.

## Other Oceanographic Information Sources

Many other sources of oceanographic information are available on the Web.

Document: Done

**http://diu.cms.udel.edu/**

Come here for information on the World Ocean Circulation Experiment and the Coupled Ocean-Atmosphere Response Experiment.

# PACIFIC MARINE ENVIRONMENTAL LABORATORY

| http://www.pmel.noaa.gov/pmelhome.html |

The Pacific Marine Environmental Lab is a division of the National Oceanic and Atmospheric Administration. And the focus of the Lab's research should be obvious from its name. Visit the Web site for more information.

# POLAR SCIENCE CENTER

| http://psc.apl.washington.edu |

The Polar Science Center (PSC) is a unit of the Applied Physics Laboratory at the University of Washington. PSC was established in 1978 at the end of the multiyear Arctic Ice Dynamics Joint Experiment (AIDJEX), a major NSF/ONR program. In 1982 PSC was incorporated into the applied Physics Laboratory, a multidisciplinary research facility.

PSC is involved in numerous studies of sea, ice, polar oceanography, and meteorology with primary funding from NASA, NOAA, NSF, and ONR.

The research focuses on observing and modeling the physical processes that control the nature and distribution of sea ice, the structure and circulation in high latitude oceans, and the interactions among air, ocean, and ice. Of particular interest is the relationship between polar regions and the global climate system. PSC employs drifting buoys and moored upward-looking sonars as well as satellite remote sensing techniques to observe ice motion and to determine ice thickness, distribution, age, and extent. Buoys and other oceanographic measurement systems are also used to study ocean dynamics and thermodynamics in ice-covered waters to gain insight into energy, exchange, and cold water processes.

Specific areas of research include:

❏ Air-ice-ocean energy transfer

❏ Fundamental theory and experiments on crystal growth

❏ Ice, ocean, and atmosphere modeling

❏ Polar climatology

❏ Polar oceanography

❏ Satellite remote sensing

❏ Sea ice buoy development

❏ Sea ice dynamics and thermodynamics

❏ Under-ice sonar and underwater vehicle development.

PSC also provides comprehensive logistics and support for numerous polar field experiments. Check out the Web site for more information.

# REGIONAL ASSOCIATION FOR RESEARCH ON THE GULF OF MAINE

**http://fundy.dartmouth.edu/rargom/**

The Regional Association for Research on the Gulf of Maine is an association of institutions that have active research interests in the Gulf of Maine and its watershed. The association was founded in 1991 and is presently housed at Dartmouth College. Members include the MIT Sea Grant Program, the University of Maine, UMass/Boston, the University of New Hampshire, and the Woods Hole Oceanographic Institution.

The basic missions of the association are to advocate and facilitate a coherent program of regional research; to promote scientific quality, and to provide a communication vehicle among scientists and the public. Its specific objectives include the following:

- ❏ To coordinate marine research and monitoring in the Gulf of Maine in order to make efficient use of resources;

- ❏ To facilitate and enhance the availability of research funds and facilities to marine scientists at member institutions through enhanced scientific planning, inter-institutional communication, and other means;

- ❏ To plan, organize, and implement long range, interdisciplinary research programs on the Gulf of Maine;

- ❏ To achieve operational economies through pooled inventories of costly supplies, standardization of major equipment items and related documentation, joint planning of equipment acquisition, and sharing of technical support personnel;

- ❏ To communicate the need for and results of scientific research on the Gulf of Maine to the user community and the public at large;

- ❏ To provide scientific and technical advice and planning for federal, regional, state, and local agencies and organizations.

# ROSENTHAL SCHOOL OF MARINE AND ATMOSPHERIC SCIENCE

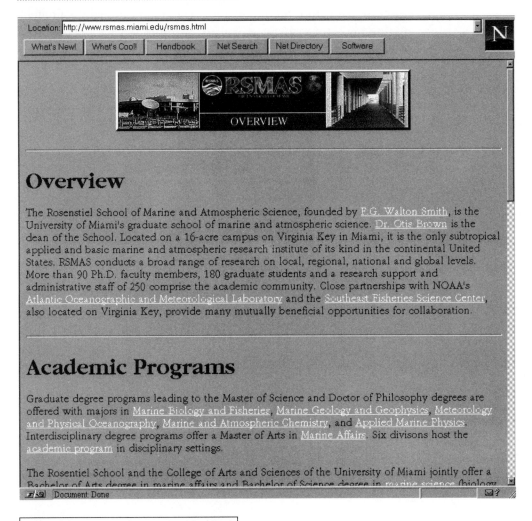

OVERVIEW

## Overview

The Rosenstiel School of Marine and Atmospheric Science, founded by P.G. Walton Smith, is the University of Miami's graduate school of marine and atmospheric science. Dr. Otis Brown is the dean of the School. Located on a 16-acre campus on Virginia Key in Miami, it is the only subtropical applied and basic marine and atmospheric research institute of its kind in the continental United States. RSMAS conducts a broad range of research on local, regional, national and global levels. More than 90 Ph.D. faculty members, 180 graduate students and a research support and administrative staff of 250 comprise the academic community. Close partnerships with NOAA's Atlantic Oceanographic and Meteorological Laboratory and the Southeast Fisheries Science Center, also located on Virginia Key, provide many mutually beneficial opportunities for collaboration.

## Academic Programs

Graduate degree programs leading to the Master of Science and Doctor of Philosophy degrees are offered with majors in Marine Biology and Fisheries, Marine Geology and Geophysics, Meteorology and Physical Oceanography, Marine and Atmospheric Chemistry, and Applied Marine Physics. Interdisciplinary degree programs offer a Master of Arts in Marine Affairs. Six divisons host the academic program in disciplinary settings.

The Rosentiel School and the College of Arts and Sciences of the University of Miami jointly offer a Bachelor of Arts degree in marine affairs and Bachelor of Science degree in marine science (biology

Document: Done

## http://www.rsmass.miami.edu/

The Rosenthal School of Marine and Atmospheric Science (RSMAS) is a graduate school of the University of Miami. Located on a sixteen-acre campus on Virginia Key in Miami, it is the only subtropical applied and basic marine and atmospheric research institute of its kinds in the continental United States. RSMAS conducts a broad range of research on local, regional, national, and global levels. Research interests pursued here include satellite oceanography (with emphasis on remote sensing and satellite imagery), an experimental fish hatchery, marine and atmospheric chemistry, comprehensive oceanic and atmospheric numerical modeling programs, sedimentary geology and marine geophysics, and ocean acoustics.

# RUTGERS INSTITUTE OF MARINE AND COASTAL SCIENCES

http://marine.rutgers.edu

Get the scoop on the latest research on marine remote sensing and physical oceanography being done at the Rutgers campus of the State University of New Jersey.

# SCRIPPS INSTITUTION OF OCEANOGRAPHY

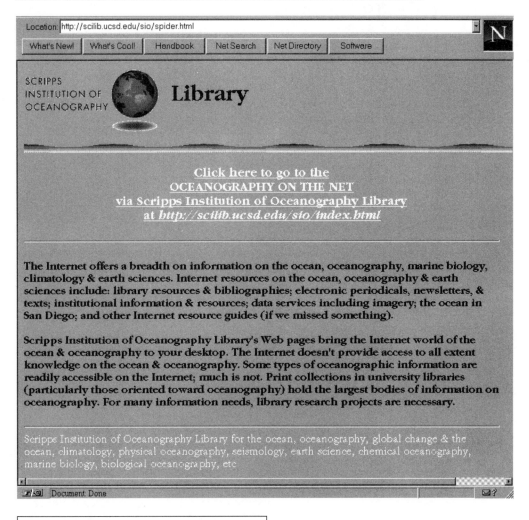

Location: http://scilib.ucsd.edu/sio/spider.html

| What's New! | What's Cool! | Handbook | Net Search | Net Directory | Software |

SCRIPPS INSTITUTION OF OCEANOGRAPHY **Library**

**Click here to go to the**
**OCEANOGRAPHY ON THE NET**
**via Scripps Institution of Oceanography Library**
**at *http://scilib.ucsd.edu/sio/index.html***

The Internet offers a breadth on information on the ocean, oceanography, marine biology, climatology & earth sciences. Internet resources on the ocean, oceanography & earth sciences include: library resources & bibliographies; electronic periodicals, newsletters, & texts; institutional information & resources; data services including imagery; the ocean in San Diego; and other Internet resource guides (if we missed something).

Scripps Institution of Oceanography Library's Web pages bring the Internet world of the ocean & oceanography to your desktop. The Internet doesn't provide access to all extent knowledge on the ocean & oceanography. Some types of oceanographic information are readily accessible on the Internet; much is not. Print collections in university libraries (particularly those oriented toward oceanography) hold the largest bodies of information on oceanography. For many information needs, library research projects are necessary.

Scripps Institution of Oceanography Library for the ocean, oceanography, global change & the ocean, climatology, physical oceanography, seismology, earth science, chemical oceanography, marine biology, biological oceanography, etc

Document: Done

## http://scilib.ucsd.edu/sio/spider.html

Scripps has put together a great metasite with hundreds of links. This is the Grand Central Station of oceanographic information on the Web.

META SITE

# SOUTH AFRICAN DATA CENTER FOR OCEANOGRAPHY

http://fred.csir.co.za/ematek/sadco/sadco.html

The South African Data Center for Oceanography (SADCO) stores, retrieves, and manipulates multidisciplinary marine information from the areas around Southern Africa. SADCO receives and stores data from the entire South African coastline, as well as the wider Atlantic, Indian, and Southern oceans in the area 0° to 70° S and 30° W to 70° E. Data sets are also obtained from other international data sources by exchange or purchase. The database contains observations since 1850, which include oceanographic station data for surface and serial depths, as well as values of temperature, salinity, sound velocity, oxygen, and nutrients. The archive also includes digitized bathythermograph and XBT data, along with surface data from voluntary observing ships including waves, wind, and weather. There are more than three million readings in the SADCO database. You can access them all via this Web site.

# SPACE OCEANOGRAPHY GROUP AT JOHNS HOPKINS

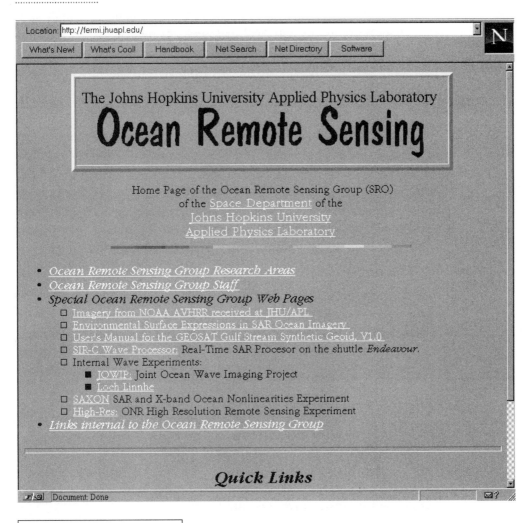

http://fermi.jhuap.edu/

The research projects conducted by the Space Oceanography Group at Johns Hopkins address problems associated with civilian and military applications of remote sensing technology in the marine environment.

The group is engaged in fundamental research on the scattering of electromagnetic energy at microwave frequencies by ocean waves to improve the ability to interpret signals from microwave sensors such as radar altimeters, scatterometers, and imaging radars. New models describing the statistical properties of microwave backscatter have been developed recently by the group, and validated using data from field experiments involving participants from U.S. and European research organizations.

The group's long-term efforts to understand radar signatures of surface currents culminated in a joint oceanographic experiment with Russian scientists in July 1992 to investigate radar imaging of oceanic internal waves off the east coast of the United States. Additional experiments took place in 1993 to demonstrate a new current measuring technique using a multiple-antenna imaging radar. In 1994 group scientists participated in the NASA SIR-C missions to demonstrate real-time monitoring of ocean waves to improve the accuracy of wave forecasting models.

Other projects of the group include investigating wind field measurements with scatterometers and synthetic aperture radars, radar imaging of mesoscale ocean features, and ocean color measurements with hyperspectral optical systems.

Visit the Web site for more information.

# VICTORIA UNIVERSITY CENTER FOR EARTH AND OCEAN RESEARCH

http://wikyonos.seaoar.uvic.ca/

The Center for Earth and Ocean Research (CEOR) at the University of Victoria was established in 1987 to initiate, foster, promote, and coordinate formal educational and research programs collaboratively with other institutions and agencies, especially the nearby Canadian government laboratories on Vancouver Island. The center's focus is on research into acoustical oceanography, carbon cycle modeling, climate modeling, ocean biogeochemical processes, geophysics, and ocean turbulence.

# WOODS HOLE OCEANOGRAPHIC INSTITUTION

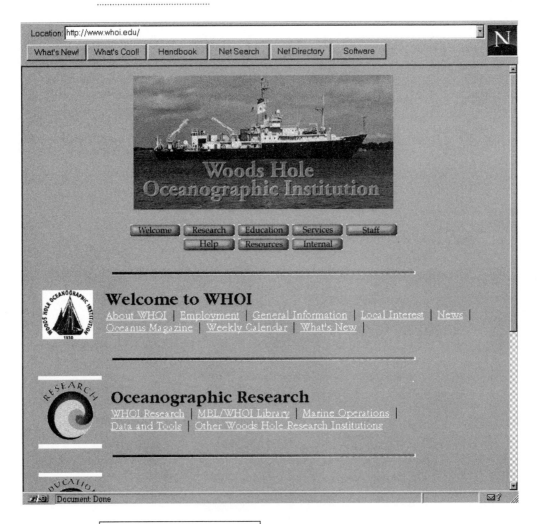

## http://www.whoi.edu/

The Woods Hole Oceanographic Institution needs no introduction. It is the largest independent marine science research facility in the United States. Founded in 1930, the institution is dedicated to the study of all aspects of marine science and the education of marine scientists. Come to this Web site for complete information on all the institution's programs.

# ORIGINS: PALEONTOLOGY AND PALEO-ANTHROPOLOGY

●–●–●–●–●–●–●–●–●–●–●–●–●–●–●–●–●–●–●–●–●–●–●–●–●–●–●–●–●–●–●–●–●–●–●–●–●–●–●–●

The **Ancient American** Magazine
Canada in the Age of Dinosaurs
The Cincinnati Museum Center
The Dinosaur Egg Project
Dinosaur Reference Center Home Page
Dinosauria On-Line
Discovery of a Paleolithic Painted Cave at Vallon-Pont-d'Arc (Ardéche)
The Field Museum On-Line: Multimedia Dinosaur Exhibit
Human Origins and Evolution in Africa
MegaLithic: Prehistoric Ritual Monuments in the British Isles
Niel's Time Lines and Scales of Measurement List
Paleonet

The Paleontological Association
Paleontological Society Homepage
Paleontology Images
Paleontology Museum, University of Michigan
Paleontology Without Walls
Plant Fossil Record Database
The (Electronic) Prehistoric Shark Museum
The Rock Shelter at Riparo Cavallino (Monte Covolo)
The Royal Tyrell Museum of Paleontology: Virtual Tour
The **Society of Vertebrate Paleontology News Bulletin**
Systematics in Prehistory: A Hypertext Edition
Talk.Origins Archive
Zachary's Paleontological Web Server

# THE ANCIENT AMERICAN MAGAZINE

http://discover.discover-net.net/ancient-american/

The purpose of the *Ancient American* is twofold: to report in layman's language the variety of ancient artifacts found in the Americas and to open a forum for discussion between the professional and the avocational archeologist. You get detailed information on the magazine at this Web site, along with late-breaking archaeological news and pictures from recent archaeological finds.

# CANADA IN THE AGE OF DINOSAURS

http://oberon.educ.sfu.ca/Dino/table.html

Dinosaurs and other ancient reptiles once roamed over the land mass that would become Canada, swam in the inland seas that once covered the Canadian prairies, and glided over warm regions that are now ice and tundra.

Come to this Web site and point and click your way to information on:

❏ the great Canadian dinosaurs—with emphasis on the four animals that typify the giant reptiles of Canada: the long-necked Massospondylus; Platecarpus, a fierce sea reptile; the horned Styrachosaurus; and the ferocious Albertosaurus;

❏ the changing Earth—how the globe changed dramatically during the age of dinosaurs;

❏ dinosaur life—how fossil evidence has led to many discoveries about the behavior of dinosaurs;

❏ dinosaur size and shape—how recent discoveries have uncovered both the largest and the smallest Canadian dinosaurs yet, and are revealing new clues about dinosaur physiology.

# THE CINCINNATI MUSEUM CENTER

http://ucaswww.mcm.uc.edu/geology/crest/cmnh/cmc.htm

The Cincinnati Museum Center is using the Web to give paleontologists details of the museum's invertebrate and vertebrate paleontology research programs, information on the Cincinnati Fossil Festival, and an inventory of the museum's extensive, major holdings in its invertebrate paleontology collection.

# THE DINOSAUR EGG PROJECT

http://infolane.com/infolane/apunix/sci-jur.html

The Dinosaur Egg Project is interesting for several reasons. It involves examining one of the earliest living creatures with state-of-the-art modern techniques and equipment. The story of the dinosaur eggs is a fascinating one in its own right; however, equally fascinating is how the limits of technology are being pushed to try and reveal what happened hundreds of millions of years ago.

A while ago a nest of dinosaur eggs was loaned to the San Diego Natural History Museum by the Rocky Mountain Natural History Museum in Bozeman, Montana. Jack Horner, world-renowned dinosaur expert, is the curator of paleontology at the Rocky Mountain Natural History Museum. He found the nest during one of his expeditions.

The nest of eggs was found at a site in Montana now called Egg Mountain. The nest was uncovered when thirty tons of rock were removed from the mountain with jackhammers and dynamite.

It is believed that the eggs were from the dinosaur species hypsilophodontid. Skeletons from this dinosaur indicate that it was a small dinosaur reaching approximately seven feet long when it was fully grown. The hypsilophodontid is a member of the duck-billed dinosaur family. It is believed the hypsilophodontids lived in colonies similar to penguins, and like penguins they laid their eggs in clutches. Appropriately enough, Egg Mountain was once part of a shallow lake that was located approximately 200 miles from the sea.

To study the eggs, Glenn Daleo, a radiologist at the Children's Hospital in San Diego, developed a unique technique for studying fossils using a CAT scanner. The CAT scan is an excellent technique for obtaining information on fossils. CAT scans are used in medical studies because they can reveal detailed information about the interior of the human body. CAT scans allow you to view the human anatomy in multiple views called slices. The scans, as they pass through the body, are attenuated differently as a result of different parts of the body having different densities. The radiation is recorded by sensitive detectors that then yield pictures.

To get detailed information on fossils via CAT scans, Glenn Daleo invented what was later named (by Jack Horner) the "wet rag method." This involves making the CAT scanner *think* it is looking at human bodies instead of rocks, since scanners are calibrated and set to deal with the various densities in the bones and surrounding tissues of humans.

Daleo placed the fossilized eggs in several plastic garbage bags. He then put the bags in a bucket of wet towels. The bags were meant to protect the fossils since water can dissolve the sandstone that the fossil is preserved in. The wet towels were meant to serve the role of organs in the human body; and the bucket was like the skin.

The CAT scans that resulted revealed more information than had been previously obtained, but the data were limited to two-dimensional black and white images that were revealing only to radiologists. Laymen found them difficult to interpret. To solve this, Glenn attempted three-dimensional visualization using CEMAX, a sophisticated imaging package that allows doctors, technicians, and scientists to create three-dimensional images from CAT scans.

CEMAX, which runs on a Sun Microsystems workstation, is in fact the leading software for rendering volume images. Different colors are selected to correspond to different densities. The various layers of the image can be dissolved away creating very clear pictures of the object. The CEMAX software allows the image to be rotated so that the user can get different perspectives of the object.

Visit this Web site and view the fantastic archive of images that were thus created.

# DINOSAUR REFERENCE CENTER HOME PAGE

http://www.crl.com/~7Esarima/dinosaurs/

Here Stanley Friesen (sarima@netcom.com) elucidates his attempt at a classification of all dinosaurs. Friesen's approach uses the standard Linnaean system and does not attempt to follow common cladistic practice. (For a more complete description of what this means, and why Friesen is doing things this way, see his discussion of taxonomic philosophy, which you will also find here on-line.) The site also includes a cladogram (a type of evolutionary tree) for the dinosaurs.

# DINOSAURIA ON-LINE

http://www.dinosauria.com/

This commercial site features the Dino Store (buy fossil replicas, dinosaur books, collectibles, etc., on-line) as well as:

- the *Online Journal of Dinosaur Paleontology*—featuring articles and discussions from enthusiasts and paleontologists on various dinosaur topics;

- the *Online Dinosaur Omnipedia*—anatomical and paleontological terms, a name-translation dictionary, lists of dinosaur genera, cladograms, and the names and dates of all the geological time periods;

- hot links to other dinosaur and paleontological sites on the Web.

# DISCOVERY OF A PALEOLITHIC PAINTED CAVE AT VALLON-PONT-D'ARC (ARDECHE)

http://www.culture.fr/culture/gvpda-en.htm

An exceptionally important archaeological discovery was recently made in the Ardéche gorges (southern France), on the edge of a national reserve, in the form of a vast underground network of caves decorated with painting and engraving dating from the Paleolithic age.

After clearing a narrow passageway at the back of a minor cave in the cliffs of the Cirque d'Estre, the discoverers made their way down a shaft and came out in a vast, totally untouched, network of caverns, full of calcite formations. Here they found several large galleries decorated in places with paintings and engraving representing animals, either isolated or organized in scenes containing over fifty at a time, the sizes varying between 0.5 and 2 m long. In total more than 200 black or red ocher paintings or engravings have been discovered to date. In several cases paintings have been superimposed and some have become encrusted with calcite.

Along the several hundred meters of galleries there are depicted a particularly large and unusual variety of animals (horses, rhinos, lions, bison, wild ox, bears, a panther, mammoths, ibex, an owl, etc.) together with symbols, panels filled with dots, and outlines of hands.

Come to this Web site to view images of the ancient artwork, and to read scholarly discussions of its meaning.

# THE FIELD MUSEUM ON-LINE: MULTIMEDIA DINOSAUR EXHIBIT

http://www.bvis.uic.edu/museum/

The Field Museum has mounted a beautifully executed on-line exhibit leveraging all the multimedia capabilities of the Web. Treat yourself to this experience.

# HUMAN ORIGINS AND EVOLUTION IN AFRICA

http://www.indiana.edu/~origins/

Come to this page for a splendid array of links and data files relating to the topic at hand: human origins and evolution in Africa.

# MEGALITHIC: PREHISTORIC RITUAL MONUMENTS IN THE BRITISH ISLES

http://rubens.anu.edu.au/mega/index.html

In 1988 Roger Martlew and Clive Ruggles of Leicester University produced an interactive videodisc containing over 2,000 images of prehistoric ritual monuments within the British Isles, "walkabouts" of sites such as Avebury, Stonehenge, and Callanhis, and several short films showing events such as midwinter sunrise at Newgrange. There were also site plans and aerial photographs. The disc was intended as a resource for teaching and learning.

Those images for which permission is available (approximately 250) have now been transferred to ArtServe and are available via this Web site.

## NIEL'S TIME LINES AND SCALES OF MEASUREMENT LIST

> http://xalph.ast.cam.ac.uk/public/niel/scales.html

Here William Nielsen Brandt (niel@ast.cam.ac.uk) provides some time lines and scales of measurement lists that he has developed. *New Scientist* Magazine (July 30, 1994, page 35) loved these time lines and scales. Brandt presents his work in both TEX and DVI files. The TEX files are in plain TEX and require no macro files. You need to have a DVI viewer hooked into your browser to look at the DVI files. The time lines include a science/technology history time line, an evolutionary/geological time line, and a cosmological time line.

## PALEONET

> http://www.nhm.ac.uk/paleonet/

Paleonet is a large collection of list servers, WWW pages, gopher holes, and ftp sites designed to enhance electronic communication among paleontologists. The site includes lots of great Macintosh and DOS/Windows software (freeware and shareware) of use in paleontology and available for free downloading.

# THE PALEONTOLOGICAL ASSOCIATION

http://www.nhm.ac.uk/paleonet/PalAss/PalAss.html

The Paleontological Association is one of the most active and prominent of the numerous societies that today cater to paleontological interests. Founded in 1957, the association aims to further the study of paleontology through publication of academic journals (*Paleontology* and *Special Papers in Paleontology*), newsletters, and a series of field guides. The members of the association encompass all aspects of paleontology, including macropaleontology, micropaleontology, paleobotany, vertebrate paleontology, paleoecology, and biostrategraphy. Come to this Web site to get more information on the association, and to access a great searchable list of paleontologists' e-mail addresses worldwide.

# PALEONTOLOGICAL SOCIETY HOMEPAGE

http://www.uic.edu/orgs/paleo/homepage.html

Come here for complete information on the society, including details on their publications including the *Journal of Paleontology*.

# PALEONTOLOGY IMAGES

http://www.nhm.ac.uk/paleonet/Visions/Visions.html

Paleontology is a highly image-oriented field. Paleontologists use images of fossils, maps, charts, and diagrams because this information cannot be adequately described in any other manner. This image-dependence has always been something of a hindrance for paleontologists because good images have traditionally taken inordinate amounts of time to create and were very expensive to publish. However, digital imaging technology has helped enormously in overcoming these problems. Come to this engaging Web page for details on all the tools and techniques that provide paleontologists with a much greater degree of control over the image they need to accomplish for their work. Here you will also find extensive image archives.

# PALEONTOLOGY MUSEUM, UNIVERSITY OF MICHIGAN

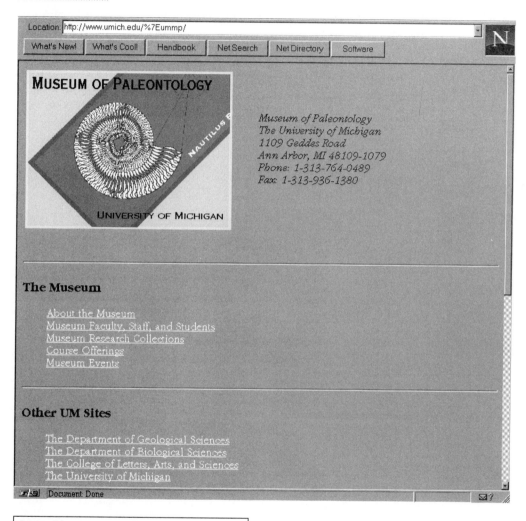

Location: http://www.umich.edu/%7Eummp/

What's New! | What's Cool! | Handbook | Net Search | Net Directory | Software

**MUSEUM OF PALEONTOLOGY**

NAUTILUS

UNIVERSITY OF MICHIGAN

Museum of Paleontology
The University of Michigan
1109 Geddes Road
Ann Arbor, MI 48109-1079
Phone: 1-313-764-0489
Fax: 1-313-936-1380

**The Museum**

About the Museum
Museum Faculty, Staff, and Students
Museum Research Collections
Course Offerings
Museum Events

**Other UM Sites**

The Department of Geological Sciences
The Department of Biological Sciences
The College of Letters, Arts, and Sciences
The University of Michigan

Document Done

**http://www.umich.edu/%7ummp/**

The Museum of Paleontology of the University of Michigan is an internationally recognized repository for collections of fossil specimens. According to the Paleontological Society and the Society of Vertebrate Paleontology, the museum ranks fourth nationally among university museums in the size of its invertebrate and vertebrate collections. In addition to research collections, the museum has a large laboratory for fossil preparation and casting, an excellent morphometrics lab, and a publication series for the dissemination of paleontological research. Visit the Web site for more information.

# PALEONTOLOGY WITHOUT WALLS

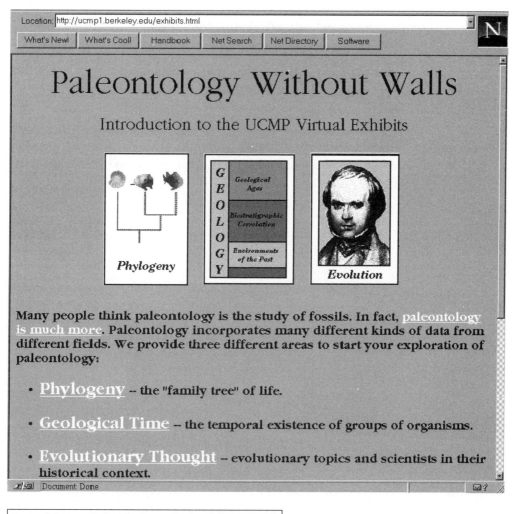

Location: http://ucmp1.berkeley.edu/exhibits.html

What's New! | What's Cool! | Handbook | Net Search | Net Directory | Software

## Paleontology Without Walls

### Introduction to the UCMP Virtual Exhibits

*Phylogeny*

GEOLOGY — Geological Ages · Biostratigraphic Correlation · Environments of the Past

*Evolution*

Many people think paleontology is the study of fossils. In fact, paleontology is much more. Paleontology incorporates many different kinds of data from different fields. We provide three different areas to start your exploration of paleontology:

- **Phylogeny** -- the "family tree" of life.

- **Geological Time** -- the temporal existence of groups of organisms.

- **Evolutionary Thought** -- evolutionary topics and scientists in their historical context.

Document: Done

**http://ucmp1.berkeley.edu/exhibits.html**

Paleontology Without Walls comprises a fascinating hypertext multi-media presentation of all facets of the science of paleontology.

# PLANT FOSSIL RECORD DATABASE

http://sunrae.vel.ac.uk/palaeo/index.html

The International Organization of Paleobotany (IOP) manages this database featuring descriptive details of most plant fossil genera and of those modern genera that have fossil species. The database also provides records of some fossil occurrences taken from the published literature and museum catalogues. The occurrences provide geographical and stratigraphical information as well as the name of the author.

# THE (ELECTRONIC) PREHISTORIC SHARK MUSEUM

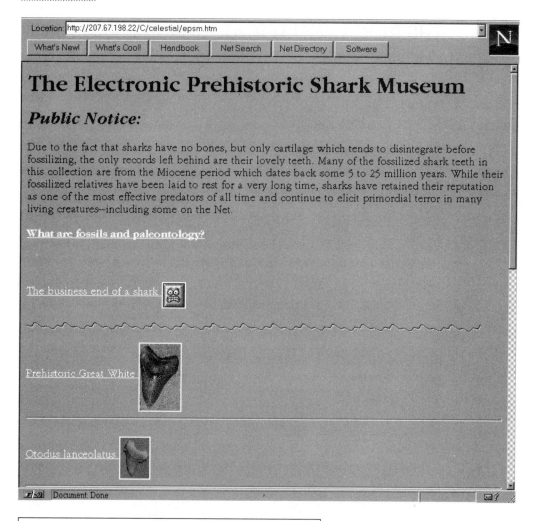

**The Electronic Prehistoric Shark Museum**

***Public Notice:***

Due to the fact that sharks have no bones, but only cartilage which tends to disintegrate before fossilizing, the only records left behind are their lovely teeth. Many of the fossilized shark teeth in this collection are from the Miocene period which dates back some 5 to 25 million years. While their fossilized relatives have been laid to rest for a very long time, sharks have retained their reputation as one of the most effective predators of all time and continue to elicit primordial terror in many living creatures--including some on the Net.

What are fossils and paleontology?

The business end of a shark

Prehistoric Great White

Otodus lanceolatus

**http://206.126.103.22/C/celestial/epsm.htm**

Due to the fact that sharks have no bones, but only cartilage that tends to disintegrate before fossilizing, the only records left behind are their teeth. Come to this Web page for a superb image archive of fossilized shark teeth from the Miocene period, which dates back some 5 to 25 million years.

# THE ROCK SHELTER AT RIPARO CAVALLINO (MONTE COVOLO)

http://www.bham.ac.uk/BUFAU/Projects/MC/intro.html

The rock shelter and prehistoric burial site of Riparo Cavallino is situated on the hill of Monte Covolo, which lies to the west of Lake Garda in Northern Italy. Monte Covolo is one of the first limestone hills of the Brescian pre-Alps. It is conical in shape, rises to a peak of 554 meters above sea level, and towers 300 meters above the surrounding countryside. The findings at Riparo Cavallino are key to our understanding of the transition from the Neolithic to the Early Bronze Age in Northern Italy. Come to this Web page for detailed reports and great images regarding the shelter and its significance.

# THE ROYAL TYRELL MUSEUM OF PALEONTOLOGY: VIRTUAL TOUR

http://www.tyrell.com/

Take a virtual tour of the Royal Tyrell Museum of Paleontology. It is almost as good as being there!

# THE SOCIETY OF VERTEBRATE
# PALEONTOLOGY NEWS BULLETIN

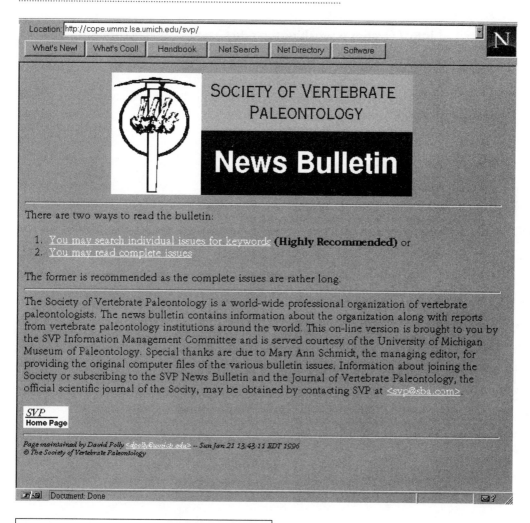

There are two ways to read the bulletin:

1. You may search individual issues for keywords **(Highly Recommended)** or
2. You may read complete issues

The former is recommended as the complete issues are rather long.

The Society of Vertebrate Paleontology is a world-wide professional organization of vertebrate paleontologists. The news bulletin contains information about the organization along with reports from vertebrate paleontology institutions around the world. This on-line version is brought to you by the SVP Information Management Committee and is served courtesy of the University of Michigan Museum of Paleontology. Special thanks are due to Mary Ann Schmidt, the managing editor, for providing the original computer files of the various bulletin issues. Information about joining the Society or subscribing to the SVP News Bulletin and the Journal of Vertebrate Paleontology, the official scientific journal of the Socity, may be obtained by contacting SVP at <svp@sba.com>.

*SVP*
**Home Page**

*Page maintained by David Polly <dpolly@umich.edu> -- Sun Jan 21 13:43:11 EDT 1996*
*© The Society of Vertebrate Paleontology*

Document: Done

## http://cope.ummz.lsa.umich.edu/svp/

The Society of Vertebrate Paleontology is a worldwide professional organization of vertebrate paleontologists. Their on-line news bulletin contains information about the organization along with reports from vertebrate paleontology institutions around the world.

# SYSTEMATICS IN PREHISTORY: A HYPERTEXT EDITION

| http://weber.u.washington.edu/~anthro/BOOK/book.html |

Here is a complete hypertext edition of Robert C. Dunnell's classic book, *Systematics in Prehistory*. The book is easy to search and navigate, and access is free.

# TALK.ORIGINS ARCHIVE

| http://rumba.ics.uci.edu:8080/ |

Talk.Origins is an Internet newsgoup devoted to the discussion of issues related to biological and physical origins. Topics discussed include evolution, creation, abiogenesis, catastrophism, cosmology, and theology. At this Web site you will find an archive of lively exchanges participated in by members of the newsgroup, as well as details on how to subscribe (for free) to the newsgroup yourself.

# ZACHARY'S PALEONTOLOGICAL WEB SERVER

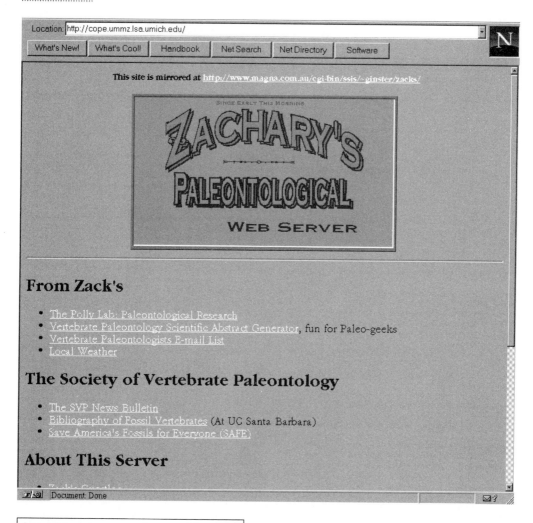

Location: http://cope.ummz.lsa.umich.edu/

| What's New! | What's Cool! | Handbook | Net Search | Net Directory | Software |

This site is mirrored at http://www.magna.com.au/cgi-bin/ssis/~ginster/zacks/

SINCE EARLY THIS MORNING

**ZACHARY'S PALEONTOLOGICAL WEB SERVER**

## From Zack's

- The Polly Lab: Paleontological Research
- Vertebrate Paleontology Scientific Abstract Generator, fun for Paleo-geeks
- Vertebrate Paleontologists E-mail List
- Local Weather

## The Society of Vertebrate Paleontology

- The SVP News Bulletin
- Bibliography of Fossil Vertebrates (At UC Santa Barbara)
- Save America's Fossils for Everyone (SAFE)

## About This Server

Document: Done

**http://cope.ummz.lsa.umich.edu/**

Here is one more paleontology metasite packed with links to interesting information. I have no idea who Zachary is, but he has assembled an excellent set of links.

META SITE

# PHYSICS

American Institute of Physics Electronic Newsletters
American Journal of Physics
Applied Physics Center
Atomic Physics Links
Brown University Physics News
Carnegie-Mellon University Department of Physics
Center for Atomic-Scale Materials Physics, Technical
     University of Denmark
The European Physics Society
Fermi National Accelerator Laboratory
Freeware for Atomic and Plasma Physics
General Relativity Servers Around the World

Institute of Physics Electronic Journals
MIT Center for Theoretical Physics
Nobel Prizes in Physics
Wolfgang Pauli (1900–1958)
**Physical Review** On-Line
Physics Around the World
Physics Lovers' Paradise (History of Physics)
**Physics Today** Magazine
Leo Szilard Home Page
UK Physics Servers
Yale: Physics, Applied Physics, and Astrophysics

# AMERICAN INSTITUTE OF PHYSICS ELECTRONIC NEWSLETTERS

http://www.aip.org/aip/enews.html

The American Institute of Physics (AIP) currently makes four of its newsletters available over the Web. These are *FYI: The American Institute of Physics Bulletin of Science Policy News, Physics News Update, Physics Education News,* and the *Center for History of Physics Newsletter.* In addition, the AIP also makes available here its *Electronic Publishing News,* a newsletter highlighting the efforts of the AIP and its member societies in the rapidly evolving area of electronic publishing.

# AMERICAN JOURNAL OF PHYSICS

http://www.amherst.edu/~ajp/

Get a free read of memorable papers that have appeared in the *American Journal of Physics* through the years.

# APPLIED PHYSICS CENTER

http://apc.pnl.gov:2080

Housed at the Pacific Northwest Federal Laboratory in Richland, Washington, staff members in the Applied Physics Center combine expertise in physics, engineering, computer science and applied mathematics. Current projects include the development of customized measurement and information systems for visualizing data, implementing control processes, evaluating materials, and providing decision support analysis.

# ATOMIC PHYSICS LINKS

Atom Trainers at the University of Wisconsin

**http://www-atoms.physics.wisc.edu**

Atomic Collisions and Plasmas at TU Wien (Austria)

**http://eaps1.tuwien.ac.at/www/atomic**

Atomic, Molecular, and Optical Physics at the University of Manitoba

**http://www.umanitoba.ca/physics/programs/atomic.html**

Atomic, Molecular, and Optical Physics at the University of Michigan

**http://gomex.physics.lsa.umich.edu/~denison/**

Atomic and Molecular Physics Division of the Harvard-Smithsonian Center for Astrophysics

**http://cfa-www.harvard.edu/~plsmith**

Atomic and Molecular Physics Group at the University of Southern California

**http://www.usc.edu/dept/p[hysics/Brochure/ AtomicPhysics.html**

Atomic Physics Division at Gaithesburg

**http://physics.mist.gov/MajResFac/EBIT/ebit.html**

Atomic Physics at Lund University, Sweden

**http://mleesun.phys.lsu.edu/dept/graduate/gatom.html**

Atomic Physics at Penn State

**http://www.phys.psu.edu/**

Atomic Physics Research, University of Aarhus, Denmark

**http://www.dfi.aau.dk/pub/res/atom/atom_phys.html**

Atomic Radioactive Physics at Auburn University

**http://www.physics.auburn.edu/atomic.html**

Atomic Spectroscopy at Lund University, Sweden

**http://www.fysik.lu.se/home/dept/spek.txt**

Atomic Structure and Calculations at Vanderbilt

**http://www.vuse.vanderbilt.edu/~eff/mchf.html**

Centre for Atomic, Molecular, and Surface Physics at Murdoch University, Australia

**http://physpc8.murdoch.edu.au/camsp**

Chemical Physics Program at the University of Nevada, Reno

**http://www.unr/index.html**

Continuous States of Atoms and Molecules at the Daresbury Laboratory, England

**http://www.dl.ac.uk/CCP/CCP2/main.html**

European Network for Highly Charged Ion-Surface Interactions

**http://kviexp.kvi.nl/disk$1/atoompc/www_serv/network.html**

Experimental AMO Physics and Quantum Electronics at SUNY, Stony Brook

**http://www.sunysb.edu/**

KVI Atomic Physics Facility at Groningen, the Netherlands

**http://kviexp.kvi.n./disk$1/atoompc/www_serv/**
**homepage.html**

Laboratory for Astronomy and Solar Physics at NASA/Goddard

**http://stars.gsfc.nasa.gov/www/welcome.html**

Laboratory of Atomic and Solid State Physics at Cornell University

**http://www.lassp.cornell.edu/**

Lawrence Berkeley Laboratory

**http://www.lbl.gov**

Los Alamos National Laboratory

**http://www.lanl.gov/**

J.R. MacDonald Lab at Kansas State University

**http://www.phys.ksu.edu/vneedham/Net/HTML/jrmlab.html**

Physics Division at Oak Ridge National Laboratory (ORNL)

**http://www.phy.ornl.gov/**

Theoretical Atomic and Molecular Physics and Astrophysics Group, University College, London

**http://jonny.phys.ucl.ac.uk/home.html**

X-Ray WWW Server at Uppsala University, Sweden

**http://xray.uu.se**

# BROWN UNIVERSITY PHYSICS NEWS

**http://www.het.brown.edu/news/index.html**

Get the latest in news of physics from around the world. The good folks at Brown Univeristy update this site daily.

# CARNEGIE-MELLON UNIVERSITY DEPARTMENT OF PHYSICS

**http://info.phys.cmu.edu/**

Visit this elegant Web page for completely details on Carnegie-Mellon's physics faculty and research programs, including exhaustive information on Carnegie-Mellon research into experimental astrophysics, theoretical and experimental high-energy physics, and experimental medium-energy physics.

## CENTER FOR ATOMIC-SCALE MATERIALS PHYSICS, TECHNICAL UNIVERSITY OF DENMARK

http://www.fysik.dtu.dk/

The Center for Atomic-Scale Materials Physics (CAMP) was founded in September 1993 by the Danish National Research Foundation to carry out research related to the fundamental understanding of the properties of materials. The research within the center is mainly concentrated in three areas: (1) the growth of materials, (2) nanostructures and interfaces, and (3) the chemical properties of surfaces. Visit this Web site for complete details on CAMP's research programs.

## THE EUROPEAN PHYSICS SOCIETY

http://www.nikhef.nl/www/pub/eps/eps.html

The home page for the leading society of Physics scholars and researchers on the European continent.

# FERMI NATIONAL ACCELERATOR LABORATORY

http://www.fnal.gov/

Fermilab is a high-energy physics laboratory, home of the world's most powerful particle accelerator, the Tevatron. Scientists from across the United States and around the world use Fermilab's resources in experiments to explore the most fundamental particles and forces of nature. Come to this Web site for reports on the latest research originating at Fermilab.

# FREEWARE FOR ATOMIC AND PLASMA PHYSICS

http://plasma-gate.weizmann.ac.il/FSfAPP.html

Courtesy of the Weizmann Institute Plasma Laboratory, download some great software that includes a symbol calculator, R.D. Cowan's LANL atomic structure software (now available in a version for DOS!), software for evaluating x-ray spectral data and photoionized plasma, a robust x-ray subroutine library, educational software for the visualization of space plasma processes, and much more.

# GENERAL RELATIVITY SERVERS AROUND THE WORLD

Cardiff Relativity Group, University of Wales College at Cardiff

http://www.astro.cf.ac.uk/groups/relativity

Center for Gravitational Physics and Geometry, Penn State
**http://www.astro.psu.edu/users/nr/**

Center for Relativity at the University of Texas, Austin
**http://godel.ph.utexas.edu/**

Imperial College, London, Physics Department
**http://euclid.tp.ph.ic.ac.uk**

Mathematical Relativity at the Australian National University
**http://einstein.anu.edu.au/**

Montana State University Physics Department
**http://www.physics.montana.edu**

Pittsburgh Relativity Group, University of Pittsburgh
**http://artemis.phyast.pitt.edu**

Syracuse University Relativity Group
**http://www.phy.syr.edu/research/relativity**

Tufts Institute of Cosmology
**http://cosmos2.phy.tufts.edu/xbook.html**

University of British Columbia Hyperspace
**http://axion.physics.ubc.ca/hyperspace/hyperspace.html**

University of Maryland Physics Department
**http://delphi.umd.edu**

University of New Brunswick Relativity
**http://www.math.umb.ca/hyperspace/**

Washington University Relativity Group
**http://wugrav.wustl.edu/**

# INSTITUTE OF PHYSICS ELECTRONIC JOURNALS

http://www.iop.org/EJ/welcome

Access full TOCs for the complete backlist of journals published by the Institute of Physics (IOP). These include journals addressing mathematical and general physics, condensed matter, applied physics, nuclear and particle physics, bioimaging, classical and quantum gravity, and many other key topics. Here you will also find information on how to arrange for full-text access to the journals through either an individual or institutional subscription.

# MIT CENTER FOR THEORETICAL PHYSICS

http://ctpa02.mit.edu/

The Center for Theoretical Physics (CTP) is a division of the Laboratory for Nuclear Science at the Massachusetts Institute of Technology. The center consists of research groups in nuclear and particle physics. Here at the center's Web site you will find progress reports for research into nuclear theory, particle theory, along with recent papers by others researchers at the Laboratory for Nuclear Science.

# NOBEL PRIZES IN PHYSICS

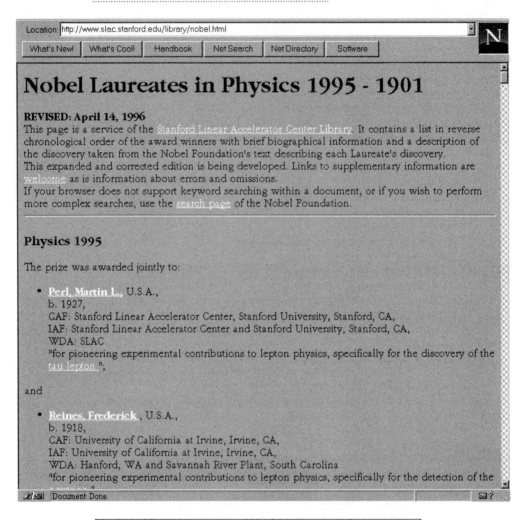

# WOLFGANG PAULI (1900-1958)

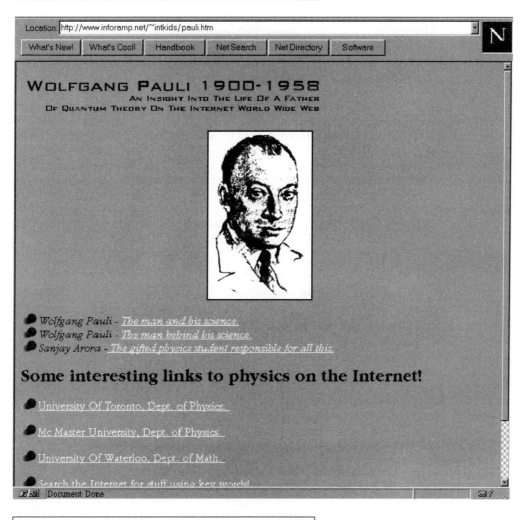

**http://www.inforamp.net/~intkids/pauli.htm**

This Web site provides complete personal and scientific biographical information on physicist Wolfgang Pauli.

# PHYSICAL REVIEW ON-LINE

http://www.c3.lanl.gov:8080/apswelcome

The home page for one of the world's leading physics publications.

## PHYSICS AROUND THE WORLD

http://tph.tuwein.ac.at/physics-services/physics_services.html

Here you have wonderful metasite access to links for just about every major physics Web resource. These include:

- ❑ Institutions and research groups—Institutions arranged by country and by field of research;

- ❑ The physics bookshelf—Preprints, journals, publishers, and on-line hypertext books (including *The Living Encyclopedia of Physics*);

- ❑ Meetings, jobs, and funding—Listings of scholarships, grants, conference calendars, and job openings;

- ❑ Educational resources and student pages—The Virtual Laboratory, physics graduate programs, student societies, and other information sources for students;

- ❑ Computing and software—Scientific computing centers for physics, and free downloadable physics software;

- ❑ Companies, instruments, and consulting—Firms involved in physics research and physics instrument development, patent information, and the Internet Market Place for Scientists (a dynamic forum for buying and selling instruments).

# PHYSICS LOVERS' PARADISE (HISTORY OF PHYSICS)

http://www.qm.ac.uk/~zgap4027/physics.html

Here is a treasure-trove of on-line information regarding the history of physics. The site features links that connect you to information on Einstein, Hans Bethe, Niels Bohr, Ludwig Boltzman, Max Born, Paul Adrien Maurice Dirac, Enrico Fermi, Richard Feynman, Werner Heisenberg, Max Planck, Frederick Reines, and many other pioneers

# PHYSICS TODAY MAGAZINE

http://www.aip.org/pt/phystoday.html

Check out the home page for the today's premier physics magazine.

# LEO SZILARD HOME PAGE

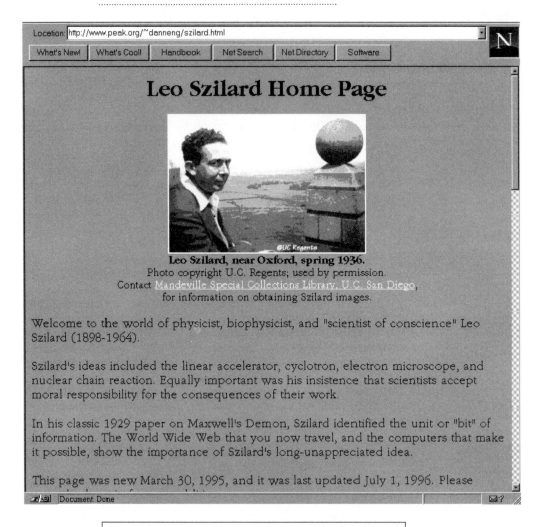

## Leo Szilard Home Page

**Leo Szilard, near Oxford, spring 1936.**
Photo copyright U.C. Regents; used by permission.
Contact Mandeville Special Collections Library, U.C. San Diego,
for information on obtaining Szilard images.

Welcome to the world of physicist, biophysicist, and "scientist of conscience" Leo Szilard (1898-1964).

Szilard's ideas included the linear accelerator, cyclotron, electron microscope, and nuclear chain reaction. Equally important was his insistence that scientists accept moral responsibility for the consequences of their work.

In his classic 1929 paper on Maxwell's Demon, Szilard identified the unit or "bit" of information. The World Wide Web that you now travel, and the computers that make it possible, show the importance of Szilard's long-unappreciated idea.

This page was new March 30, 1995, and it was last updated July 1, 1996. Please

Document: Done

## http://www.peak.org/~danneng/szilard.html

Visit that section of cyberspace where physicist, biophysicist, and "scientist of conscience" Leo Szilard (1898–1964) lives again. Szilard's ideas included the linear accelerator, cyclotron, electron microscope, information theory, and the nuclear chain reaction. Equally important was his insistence that scientists accept moral responsibility for the consequences of their work. In his classic 1929 article on Maxwell's Demon, Szilard identified the unit of "bit" of information. The World Wide Web and the computers that make it possible show the importance of Szilard's long-unappreciated idea.

# UK PHYSICS SERVERS

http://euler.ph.ic.ac.uk/UKPhys.Serv.html

Here you will find a long and complete list of links to virtually (no pun intended) all the physics Web sites in the United Kingdom, including labs at Aberdeen University, Birmingham University, Bradford University, Bristol University, Cambridge University, Durham University, Edinburgh University, Heriot-Watt University, Hull University, Leicester University, and many other institutions.

The Web document includes a biographical chronology of Szilard, the text of an interview with Szilard about President Truman and the atomic bomb, the text of an FBI interview with Albert Einstein concerning Szilard's security clearance, a wonderful 1935 photo of Szilard with Ernest O. Lawrence, and a large archive of papers related to the decision to use the first atomic bombs (and Szilard's attempts to prevent this). This archive includes a 38K image of Leo Szilard's historic petition to President Truman protesting the use of the atomic bomb. (Your monitor must support at least 800 × 600 resolution to view it properly.)

The site also provides links to additional archives relating to Hiroshima and Nagasaki, and a link to the Register of the Leo Szilard Papers at the Univeristy of California, San Diego. (Warning! The file is 172K! Be armed with either a fast connection or be prepared to wait.)

# YALE: PHYSICS, APPLIED PHYSICS, AND ASTROPHYSICS

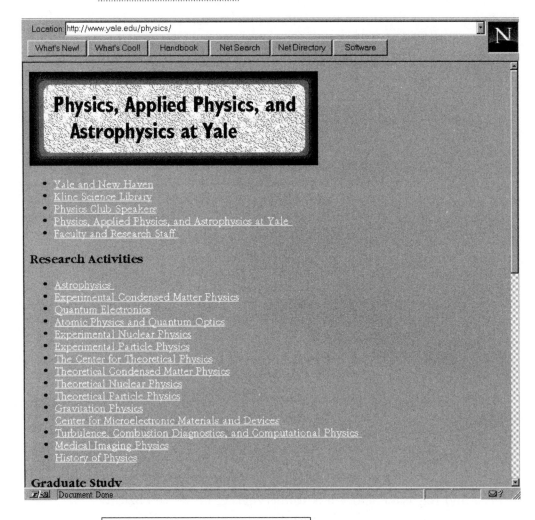

Location: http://www.yale.edu/physics/

What's New! | What's Cool! | Handbook | Net Search | Net Directory | Software

## Physics, Applied Physics, and Astrophysics at Yale

- Yale and New Haven
- Kline Science Library
- Physics Club Speakers
- Physics, Applied Physics, and Astrophysics at Yale
- Faculty and Research Staff

**Research Activities**

- Astrophysics
- Experimental Condensed Matter Physics
- Quantum Electronics
- Atomic Physics and Quantum Optics
- Experimental Nuclear Physics
- Experimental Particle Physics
- The Center for Theoretical Physics
- Theoretical Condensed Matter Physics
- Theoretical Nuclear Physics
- Theoretical Particle Physics
- Gravitation Physics
- Center for Microelectronic Materials and Devices
- Turbulence, Combustion Diagnostics, and Computational Physics
- Medical Imaging Physics
- History of Physics

**Graduate Study**

Document: Done

## http://www.yale.edu/physics/

Get the story on the latest Yale research in astrophysics, experimental condensed matter physics, quantum electronics, atomic physics and quantum optics, experimental nuclear physics, experimental particle physics, theoretical condensed matter physics, theoretical nuclear physics, theoretical particle physics, gravitation physics, and more.

# PUBLISHERS & BOOKSELLERS

Last but not least, here is a quick list of scientific publishers and booksellers who have put homepages, catalogues, and sometimes even portions of books and journals on to the Web for your perusal.

Academic Press

**http://www.hbuk.co.uk/ap/apwelcom.hem**

Addison Wesley

**http://www.aw.com/**

American Association of University Presses

**http://aaup.pupress.princeton.edu**

The American Chemical Society

**gopher://acs.infor.com:4500/**

American Institute of Physics

**http://www.aip.org/aip/aipress.html**

American Mathematical Society

**http://www.ams.org**

Artech House

**gopher://gopher.std.com/11/Book%20Sellers/artech**

The Astronomy Book Club

**http://www.booksonline.com**

Baltzer Science Publishers

**http://www.NL.net/~baltzer**

Barnes & Noble

**http://www.barnesandnoble.com**

BIOS Scientific Press

**http:/www.demon.co.uk/bookshop/bicat.html**

Blackwell Scientific

**http://www.blacksci.co.uk**

The BookPool—Discounted Technical Books

**http://www.bookpool.com/**

Boyd & Fraser

**http://www.bf.com/bf.html**

Cambridge University Press

**http://www.cup.cam.ac.uk/**

Cold Spring Harbor Laboratory Press

**http://www.cshl.org/about_cshl_press.html**

CRC Press

**http://www.crc.press.com/**

Data Works Express Technical Books

**http://www.dataexpress.com/**

Duke University Press
**http://www.duke.com/DukePress**

Ecola's Technical Book Directory
**http://www.ecola.com/ez/books.htm**

Elsevier Science
**http://www.elsevier.nl/**

Gordon & Breach
**http://www.twics.com/%7Egbtokyo/home.html**

Internet Book Shop
**http://www.bookshop.co.uk/**

The Library of Science Book Club
**http://www.booksonline.com**

MIT Press
**http://mitpress.mit.edu**

The Natural Science Book Club
**http://www.booksonline.com**

Oxford University Press
**http://www.oup.co.uk/**

Protech Books Bookseller
**http://www.pro-tech.com**

Springer Verlag—Berlin/Heidelberg
**http://tick.ntp.springer.de**

Springer Verlag—New York
**http://www.springer-ny.com**

Taylor & Francis

**http://www.tandf.co.uk/**

The Virtual Bookshop

**http://www.virtual.bookshop.com**

John Wiley & Sons, Inc.

**http://www.wiley.com**

# INDEX